To Cathy Shelton

RIVERSIDE

Miriam Poe Ryan

RIVERSIDE

Historical Novel

The incredible story of
Marcus Daly, founder of
Anaconda Copper, pioneer and
citizen of Montana

by

Miriam Poe Ryan

Vincent Publishing
P.O. Box 1258
Hamilton, Montana 59840
406-363-2291

RIVERSIDE

Historical Novel

The incredible story of Marcus Daly,
founder of Anaconda Copper

BY

©Miriam Poe Ryan

First edition published in 1994
Revised edition 1995

ISBN 0-9645678-0-6

Printed and bound by
Signal Graphics Printing
7150 Leetsdale Drive
Denver, Colorado 80224

Published by

Vincent Publishing
Hamilton, Montana

RIVERSIDE

Historical Novel

Evaluation by Creative Writing Department
University of Montana, Missoula, Montana

"The historical style is literate and graceful and the text is very readable. The manuscript contains excellent research and much valuable detail about the Daly family."

Acknowledgements:

My thanks to:

Vera & Bill Poe

Georgiann Dayton

Ginger Renner

Dr. James Kriley

Aaron Syme

Mary Poe

Tom Ryan

Betty Lee & Powell Wilson

Prologue

The cars slowly turn off the highway into the drive lined with sugar maples glowing in the sun, colored red by the first two frosts of autumn. As the tourists enter the gates to Riverside, the impression of wealth and grandeur is magnified by the massive iron gates attached to brick columns that are finalized with the hospitality symbol, a carved stone pineapple. From the brick columns the wrought iron fence extends on each side forming a semi-circle, as if arms were drawing the visitor into the drive and its mysterious mansion.

Impressive, yes, for at the end of the one-mile tree-lined drive one can glimpse the Daly mansion on the Riverside estate: a buff-colored brick building with large white classical columns, and architecture in the style of the Georgian period.

A young woman dressed in a black skirt and white blouse and wearing a straw hat steps out from the gazebo to greet the visitors scheduled for the 11 a.m. tour. She smiles and tells them what the mansion and its grounds offer. For a dollar they can explore the grounds unescorted, see the mansion from the exterior, the laundry house, the playhouse three-quarters the size of a regular house, the boat house, ice house and, the grandest sight of all, the grounds with its fifty acres of grass and one-hundred-year-old trees. Each tree is marked with the name of its species, for all were imported and thrived in the Bitterroot Valley of Montana. Brought by Marcus Daly from all areas of the United States, the trees are sugar maple, Norway maple, lindenwood, poplar, silver maple, black oak, red oak, cottonwood, Russian olive, sycamore and the McIntosh apple trees at the side of the greenhouse area. Altogether there are twenty-five varieties.

On the grounds is the "plunge" or the swimming pool. The olympic-size pool has four ladies dressing rooms and four men's dressing rooms with a shaded lanai area in between. There are tennis courts and a bowling green. It is easy to imagine the laughter from the plunge as the young people dive and splash, and the small children in their inner tubes float at the shallow end with a local boy as a life guard keeping watch. The tennis courts could have doubles on one court and a singles game on the other, while the men could be bowling on the green, handsome in their straw bowlers, white pants and striped coats. There is even room on the green for the ladies to play croquet, a lady-like game, though the competitive spirit could give rise to an invigorating time. All of this is certainly a great sight for $1.00, but the gatekeeper says the tour of the mansion and grounds will be $4.00 for adults and $2.00 for children. As the guests hesitate deciding whether to take the tour or to just walk about the grounds, the young woman encourages them to take the tour for not only is it the largest house in Montana, 24,000 square feet with twenty-four bedrooms and fifteen baths, but also the story of Marcus Daly and his family is a fascinating narrative. The guide is so sure of her advice that she asks them to return to her after the tour of the mansion and give her their opinions. Is she correct?

The parking lot is filling, and the 11 a.m. tour is about to begin. The guests are leaving their cars and proceeding on the final walk to the Mansion. Carriages and, later, cars traveled the drive where they walked. The drive ends in a circle turn-around at the mansion door. The guests and family used to arrive in carriages with spirited horses or, in later years, large chauffeur-driven cars. One can imagine the excitement of the mansion's glory days, and now the tourists enter the magic world as guests. This tour will bring vividly to them a way of life gone forever, preserved for their enjoyment, and a story of a family that was destined to make history in domestic and foreign affairs.

The 11 a.m. tour guide is dressed the same as the greeter at the gazebo. The costume is simple, done in the 1910 manner, and helps take one back to the time when the mansion was full of guests and family. The guide beckons the guests on tour to come onto the porch, for she wants to spend a few minutes explaining the grounds and the Mansion.

Now begins the magic. "Riverside," she says, "was comprised of 28,000 acres, which in turn was divided into seven units, each with a foreman, and a manager over all. The seven units were a dairy farm, grain and alfalfa acreage, nursery stock and vegetables, fruit orchards, a cattle ranch and a thoroughbred horse farm that also quartered fine draft horses and trotters. Actually, Riverside was a self-sufficient empire with the mansion as its heart and soul."

Continuing, the guide tells of the mansion as seen today: a Georgian Revival structure registered on the National Historic Registry. This 1910 Georgian Revival was the final remodeling of the original farm house built by the Chaffins in 1878. Amazingly, there are three different mansions in the one. The house and subsequent mansions were never replaced, just remodeled. Since the greeter has given the visitors maps of the grounds and a description, the mansion tour guide takes them in the front door, welcoming them to the Daly mansion. Now the story begins of Riverside and the Marcus Daly family.

It is dim in the reception hall, for the people had been

standing in the bright sunlight on the porch; but as their eyes adjust, they can see gold damask silk wall coverings, a carved marble fireplace, parquet floors banded in mahogany, Adams style settees on each side of the hall and a grand staircase with wide steps. A large landing on the curve of the staircase leads on to the second floor. The bannister spindles are white and turned in classic tradition and the rail is mahogany.

There are lace curtains at the front windows and cut glass windows on the double-door entrance. Catching one's eye are the upholstered green velvet radiator covers in the front hall -- such a practical idea, for the radiator covers provided seating and beauty at the same time. There is so little furniture to be seen in the entrance hall, which and the tour guide explains. "The Daly Mansion had been boarded up for forty-one years, fully furnished even with crystal and linens, waiting to be lived in and used. The house now belongs to the State of Montana and was opened to the public in 1986. Before 1986, the furniture was sold by the heirs at auction, and everything was purchased in a frenzy. Everyone in the area wanted something from the Daly Mansion. The house stood silently behind closed gates, barely visible, boarded and mysterious. People drove by and peered in, always wondering 'why.'"

The guests' attention is drawn to three photos and one drawing. The drawing is of the original Chaffin house. Today the original Chaffin house consists of the kitchen, the cooler room, the servants' dining room and butler's pantry, the back stairs, and two bedrooms left of the stairs and one to the right. The Chaffin house was remodeled and added to for the first Victorian Mansion. The Chaffin living room became the kitchen of the Victorian mansion, and the addition included a study, two bedrooms, a living room, a sitting room, a music room downstairs and a porch all around. Upstairs were ten more bedrooms, a third floor with nine more bedrooms, and a ballroom or billiard room. Some of the bedrooms on the third floor were for the servants and others were for the families with children and other guests. All of this was finished for the summer of 1887.

After the Chaffin house and new addition had been incorporated, and the decorators had come and gone, the Marcus Daly family moved into the first Victorian house, their summer home. Their winter home was a lush apartment at 725 Fifth Avenue in New York City.

In 1896 Margaret Daly remodeled the first Victorian into the Queen Ann style. The interior was updated with wallpaper and paint, reupholstered furniture, new carpets, and some oriental rugs. The porch on the left side was enclosed, and turrets with cranberry glass were constructed. The mansion was white frame and, with the turrets and cranberry glass, it looked like a confectioner's creation.

Upon his arrival for the summer of 1896, Mr. Daly hated the Queen Ann architecture and thought it looked like a church. He and Mrs. Daly called upon Mr. Gibson, a well known architect in Missoula, Montana. Mr. Gibson provided them with a drawing of the exterior and a floor plan that enlarged the living areas by combining some of the small, more intimate Victorian rooms to make larger ones.

The present exterior is a product of the Georgian Revival, in the manner of the Georgian Architecture of England in the 1760s. The Georgian period was inspired by the rediscovery of Pompeii, by the Greek and Roman columns and motifs that were excavated from the volcanic ash that had covered the city. The Adams Brothers in England decorated their furniture with the Pompeii motifs. The golden age of English furniture and architecture brought about cabinet-making by Adams, Hepplewhite, Chippendale and others. It was agreed that a Georgian Revival remodeling would be undertaken in the near future.

The tour guide leads the group to the Italian fireplace. This fireplace is one of seven in the house imported from Italy, and the carved white marble is from the Carrara marble quarry in Italy, the same 2000-year-old quarry from which Michelangelo obtained marble for his sculptures. The fireplace mantle also has gold-colored marble fractured with brown faults. The face of the fire box and the doric columns on the side are of this marble as well. The twenty-four-inch-high andirons are of solid bronze dipped in gold, resulting in the French decoration called "ormolu." Each andiron weighs twenty pounds, and the fire screen is in the Louis XV style with ormolu decorations.

Above the fireplaces is a portrait of Marcus Daly, founder of Anaconda Copper and of the largest copper mine in the world. He was the builder of the largest ore smelter in the world and also a number of towns, including Anaconda and Hamilton, a titan of the lumber industry, a respected breeder and owner of racing horses of world renown, a loving father, and loyal friend to his workers and business associates. The guide begins the story of this remarkable person and his family.

Chapter One

Marcus Daly was born in 1841 in Cavin County, Derrylea, Balljamesduff, Ireland, fifty-seven miles northwest of Dublin. He was born on a dreary grey day with chilling northern wind that left a white frost over all the countryside.

The Daly's small stucco house was warm from the peat stove fired high for the birth. The other six children were in the far corner of the room while their mother rested between her pains by the fire. The time was near, and the children went to the loft and into their straw beds with quilts and wool blankets for warmth. Hopefully, the baby would come soon for everyone was tired and the small amount of food they had eaten would soon leave their stomachs, and hunger pangs would return. Sleep would be a blessing in many ways.

Marcus was born without too much fuss, and his mother was happy to see such a beautiful boy to go with the three girls and three boys looking down at her from the loft. "A darling boy, children, one for you to love and to help care for. One for God to give his blessings. His name will be Marcus." The day was over, all was well, with a time to sleep after prayers and thanksgiving.

At daybreak Marcus stirred and cried out for his mother to feed him. He was lying by her side, and it was not any trouble to satisfy his hunger and hold and cuddle him. The children came tumbling down from the loft and gazed and poked the baby Marcus. Patricia, the eldest, held him for a moment. The baby was a beautiful, blue-eyed, dark-haired Irish lad. He looked very Irish, as did all of the Dalys.

The first years of Marcus Daly's life went by with terrible hunger in the land. Most of Ireland's potato crop had failed in

1845, but the Daly family was fortunate in that their crop had been only lightly touched by the blight. Nevertheless, the blight was still a threat, for it could easily spread over their potato crop, which was the main food source for the Irish and also a cash crop. There was little enough to eat, and if the potato crop failed there would be starvation over all of Ireland. The other produce on the Irish tenant farms were sold to pay their absentee English landlords who had divided the estates into smaller and smaller land parcels, making it impossible for the Irish to rise above abject poverty. Before the blight, the potato crops had been the main food of the Irish and had satisfied the English demand for Irish potatoes on their plates.

By 1850, the times were terrible, and the hope and joy of the Irish was gone. People were migrating to the United States in droves, crowded in the holds of ships, some dying on the way, but knowing that they would die of hunger if they did not try for the new land and a new life. The potato blight continued to spread, and a famine set in. The Irish were dying from starvation.

Marcus was nine years old when the potato famine ended as quickly as it had begun, without any reason or explanation for the beginning or for the end. The Irish state was still impoverished and heavily taxed by the British. To help his family and earn his keep, Marcus took work at a pig farm. Pigs stay clean if they have clean water for their wallow, and Marcus vowed if he ever had pigs he would have clean water for them. Mr. Cochran didn't care if his pigs were in mud or worse.

In 1856, when Marcus was fifteen years old, he decided to immigrate to the United States. It was hard to leave his mother and father, brothers and sisters, but he could not earn enough to help the family or even himself. It was hardest to say goodbye to his father for he felt he had let him down by not being able to contribute more to the family.

In the Irish fashion there was much crying, hugging and well-wishing among the Daly family, friends and relatives who had assembled for Marcus's farewell. Finally the music started, and everyone was singing as the tears flowed along with the ale.

While the singing was at its loudest, Marcus made his way across the room to his mother and father. He looked at them. His father was a short, strong, powerful man with black hair streaked with silver and his eyes were as blue as the corn flowers that grew outside the Daly home. Marcus knew he would need to remember well the looks and touches of his own family, for it could be the last time he would see them.

His mother was a small woman who still had a spring in her step. Her hair was pulled severely into a bun on the top of her small head. Marcus would never forget the blue eyes that were able to look into his soul. He felt a great loss at leaving this woman who, God knows how, bore seven children, cared for, schooled and loved them all. But Marcus knew in his heart that he was the one his mother loved, prodded, taught to read, and put her hopes in for a bright future.

Marcus said, "Mother, you'll be all right, you and Father, without me to help here at the farm. I'll send you any extra money I have. I just hope my leaving will make your life easier."

"Marcus, how do we know what the future brings? I know you have given a great deal of thought to your going to America. I have to believe that it is meant to be and that God is with you. You must know, I will say prayers for you each day. Be strong, my love, and use what knowledge you have to your best advantage.

We will be fine here in Ireland. I hope you will write to us of life in America."

"Thank you, Mother, for everything. I'm glad you can give me your blessing. I will be worthy of you and Father's belief in me." At that, Marcus kissed his mother on the cheek and turned to his father.

"Father, my thanks to you for your help in getting the passage for me to America. I'll always remember what a great father and person you are. I only hope I can be as good. I'll send what I can to help you and the family. I wonder what it will be like without you to help me understand life."

"Marcus, you will do well in your new life's work. You are a strong young man both physically and mentally. It won't be easy, I know, son, but I think you were born under a special star. You have always been different than the rest and I believe your destiny is in America. Here -- take this gold coin with you to America. It's not much, but maybe it will help until you can earn some money. I want you to have it."

"Oh, Father, I don't want you to do any more than you have."

"I want to, Marcus. Now take this coin."

"All right, Father, but I will return it to you as soon as I can," said Marcus.

His father reached for Marcus and embraced him, and they patted each other's shoulders while tears of love filled their eyes.

Marcus then turned to his friends and relatives and spoke to them.

"Thank you for making my passage possible. I will never forget you and hope I can repay you all a hundred-fold." With that, he left for the train station and the start of his life's adventure.

As he boarded he looked around for a familiar face and, seeing none, he continued on to his place, one by the window, for he knew he may never see his homeland again.

It was a fifty-seven-mile journey from the town of Cavan to Dublin, and the countryside was green, lush, and manicured. The small farms of thirty acres and the larger ones of one hundred acres were well cared for, as were the creamy low stucco houses and outbuildings. Raising cattle, pigs and sheep on these farms was the larger portion of the farmers' income. Farming the land for crops and for food for the animals and people kept these families working from sunup to sunset.

Ireland has an abundance of rainfall. Its location in the world is favorable for four different conditions that cause rain, the climate we described as mild, moist and changeable. This continuous moisture results in vegetation with forty different shades of green, and he saw them all on the way to Dublin.

When Marcus boarded the train a soft drizzle had begun to fall. Upon arriving in Dublin, the rain gently tapped his face as he stepped off the train. The rain, almost a mist in the balmy temperature, created what was known as a "soft day" in Ireland. Marcus was glad to have the gentle "soft day" as his last in Ireland. He would always be able to recall that touch, for nowhere is there such a "soft day" as in Ireland.

Dublin was a beige city with carts, horses, carriages, hand carts and people. Marcus did not have time to explore the city, for his ship was to leave port at 6 p.m. that day.

Marcus was quickly brought back from his reverie by being jostled in line while waiting to board ship. A young couple with a child was directly behind him. The mother, lunging to retrieve her son before he fell in the water off the gang plank, had pushed Marcus from behind. After much apologizing from the young mother, Marcus said, "Better to push me and save your boy, so not to worry at all."

Marcus hoped to see more of them on the journey, but the ship was so crowded that he wondered if he would. Perhaps he could keep them in sight. He was already lonely for the clamor and warmth of his home with the other children and mother and Father.

The oceangoing sailing ship plied the Atlantic Ocean from New York to Liverpool. These sailing ships set records for speed and there were contests between ship owners and their crews. Even in 1856 the crossing was sometimes no more that eleven days.

Marcus was looking forward to being able to explore the ship. It was a beauty with sails, and he couldn't wait to see them filled with the wind, taking him to a new life.

When 9 p.m. was getting close, all visitors were asked to leave the ship. There was such wailing and tears that Marcus forgot about his own sadness as he watched the farewells of others. He could understand their grief, for it was not known if the families would see their friends, sons, or daughters again.

The schooner was ready, putting up small sails to get out of Dublin Bay and out into the Irish sea. Marcus' berth was in the hold, where he had placed his meager belongings, but he wanted to watch from the deck as they sailed out. How agile the men were as they kept letting out more sail. He hoped he would see the mountains south of Dublin, the lights of Cork and the shores of Ireland as they headed out into the Atlantic.

He felt the thrill of the race as the schooner picked up

speed. What an amazing sensation to be skimming over the water so effortlessly! Marcus noticed the quiet power of the schooner as the ship hurried on to its destination, New York City, New York, U.S.A.

Marcus stayed on deck until it was dark and he lost sight of land. In the moonless dark there were some stars, but he felt the vastness of the sky and the ocean.

The berth that Marcus had been assigned was close to the family that he had met through the jostling on the gang plank, and Marcus was comforted. He went to them and introduced himself properly. They were the Sullivan family with one child, Sean. Marcus was at ease with them for, except for the name, they could have been kin. The passage included food and, although it was simple -- bean soup, fresh bread and an apple -- the apple was a real luxury; certainly fresh fruit was not a common item for passengers. God was good and Marcus lay down in his berth, the middle bunk, and after a few thoughts and prayers, fell asleep. The ship was making a swooshing sound as it sliced through the water. The last noise he remembered before falling asleep were the 10 p.m. bells. So ended the first day that began the Marcus Daly saga.

All of Marcus' goals and desires centered on sending money back to his poor family. Even though he had worked in Ireland, what money he made was really not even enough to help himself. Surely, America would have a job for him, and he knew American farms were profitable and provided bountiful harvests. He could always work on a farm, and he didn't need much for himself.

The Atlantic crossing is never smooth in the northern route, and this voyage was no different. One the third day a storm came without warning, and there was much scrambling by the sailors to bring in the sails and tie down all loose cargo. A rope was strung along the sides of the ship so that the men could hang on to it as they made their way from one part of the schooner to the other. The passengers were told to stay below, and it was sad and sorry night. Everyone was ill and many lost their equilibrium. The trembling families stayed in their berths, hoping they would not die, for some imagined that the awful night was just another segment in their life-long ordeals. Howling wind, plunging ship, waves crashing over the decks. It was not a hurricane but it was almost as damaging. The Captain was cursing, for there had been no indication such a storm was anywhere near the route he was taking. There was nothing to do but ride it out.

One minute the ship was crashing around, the next it seemed to bump and then there was a calm. With that, the edge of the storm passed, and it was time to hoist the sails and carry on. The captain had hoped to set a time record for his ship and his company. His archrival in shipping had set a record from Liverpool to New York (eleven and one-half days) some weeks before. The captain and passengers would be happy to arrive without damage after the pounding the storm had given the ship and crew. Next time he would break the record, for his ship was the best. This trip, however, he had no cooperation from the weather.

After the storm the days were very nice, and everyone strolled and visited on the deck. Marcus took the Sullivan's young son, Sean, on a walk around the deck. The boy's eyes were huge with wonder that the sea was so near and so vast. While they were on deck a whistle blew and a sailor called "Land ho!" They had been sailing for thirteen days -- a very respectable time to cross the

Atlantic. Thirteen days before, Ireland had been Marcus' entire world, but now here was a new land, a city of great size, and all the vast beyond. New York City would be his home for awhile.

As the schooner came into the harbor, small boats whistled and followed the ship as a welcoming committee. This was a ritual of which Marcus had never seen nor heard. What a thrill it was! Then came the breathtaking view of the New York skyline.

Marcus was worried about immigrating, for he did not have a sponsor. He hesitated to ask the Sullivans to let him come with them as if he were part of their family. But they had become good friends, and they had said they would help him. He approached Mr. Sullivan and told him that he had no one to meet him, and he was afraid that immigration would detain him and he would be deported. Mr. Sullivan had family members to meet them. He had a job, and told Marcus he could be with them through immigration and stay with them until he was able to find work. And so it was.

The ship docked and the immigration process started as soon as the passengers could exit from their quarters. It was a bewildering experience for the people were herded like cattle and left for hours to mill around and worry in a large warehouse. The immigration personnel had difficulty understanding the immigrants' names and asked for the names to be spelled. That was worse, for the Latin countries and other countries used different sounds for their letters. Many immigrant's names were shortened by the immigration officers. Marcus was thankful he had Mr. and Mrs. Sullivan to help him. Marcus took Sean's hand, and to pass the time they walked in a circle around the assigned area. Finally the Sullivans and Marcus were processed. The officers paid little attention to Marcus and directed all questions to Mr. Sullivan.

After they made their way through immigration they saw Mr. Sullivans' brother was on the other side of the barrier fence to welcome them to America. Mr. Sullivan said to his brother, "Jerry, I'm so glad to see you. I wondered if I ever would!" They embraced and turned toward Mrs. Sullivan, Sean and Marcus.

Mr. Sullivan introduced his wife and Sean and explained Marcus' predicament.

Jerry said, "Tom, surely we will take care of Marcus, too. We're glad to help you, Marcus. Take a load and we will be out of here and on our way."

Marcus had passed through immigration as a young cousin of the Sullivans and officially became their responsibility. He was relieved and so pleased that he was welcomed in such a warm manner. He would earn his keep and repay their kindness.

Life was good. Marcus helped others when he could, passing on the belief that one should share good fortune with others and help those in need.

The streets weren't "paved in gold" and he didn't expect them to be. Because Marcus could and would work at any job, he became a gardener, a telegraph boy, a leather tanner and a dock worker. He saved a dollar from every two dollars he earned. Marcus sent some money home, and in his letters to his family he told of his desire to earn enough money to get passage to California and the silver and gold mines. The streets might not be paved with gold in New York, but in California there had been some lucrative strikes, and miners were always needed.

After five years, though he regularly sent money to his family, he was able to save for passage to California. He boarded a freighter to Panama, took a train and a stage across the isthmus to the Pacific Ocean, boarded another freighter and disembarked in San Francisco. The docks were surrounded by saloons and boarding houses. The streets were populated with sailors, toughs and "ladies of the night." Marcus was able to push his way and hold his own, and found a boarding house to settle in until he could get to the mines in Placerville, California. The boarding house was at the edge of the rowdy section of San Francisco that he had been through after leaving the ship. Marcus never had any trouble settling in, for he didn't have very many belongings. Marcus did have a dream — he wanted to make an honest wage, and mines paid the very best wage known.

At the boarding house he met a fellow Irishman, Thomas Murray. Thomas had the same dream as Marcus. The two got jobs as gardeners but talked about getting to the mines in Placerville. They finally saved enough to take the stage to Placerville. They didn't have jobs, but were sure something would turn up. Even if there were no mining jobs, there were camps to work in and other jobs they could do.

Placerville was a boom town. It was a tent city with muddy streets, a boarding house, saloons and "camp followers." Thomas and Marcus were tired, hungry and dirty after the two-day stage ride to Placerville. First they needed a place to stay, second, a bath at the town's bath house, then, dinner.

The morning brought them closer to their objective, a job in the mines. Although both men were good natured, this adventure together had been a trial on their ability to have a friendship. They had passed the test and were bound in homeland, new land, a desire to work and to remain friends forever.

The morning after their arrival in Placerville, they awoke early and hurried down to the local mine employment office. All the requests for labor at the mines went through this office. Thomas and Marcus took their place in line. The local bully took exception to their being there, remarking, "There are already too many Irish. I'll eliminate two." Marcus ignored his remarks, for he had heard worse said of the Irish in New York. However, the bully would not let up, so Marcus turned and hit the bully on the jaw and felled him with a knock-out blow. Marcus was hired, for this episode had not been wasted on the hiring agent. The agent was familiar with the bully and was pleased Marcus had put him in his place — flat on his back. Marcus pushed Thomas up to the hiring agent and said they had come together to find work in the mines. Thus, Marcus and Thomas became miners together.

While the other miners spent their wages and were generally out for a good time, Marcus was interested in learning the mining business and studying the geology of the mines. He noticed that certain formations would prove to have gold or silver. He learned the hard rock method of mining. He learned about sluicing. Marcus was like a sponge, absorbing all of this knowledge which later would bring about phenomenal results.

Since the town's social life centered around a saloon and the boarding house, Marcus entered into the rough miners' life in Placerville. Marcus was a man's man. He made friends easily, and he had a circle of friends who liked his work, his abilities and his stories. All Irishmen tell stories and probably would get away with some of their tales if they didn't step over the line to the ridiculous! The ladies liked Marcus, and he kidded with them, but never seemed to need them. Actually, there was little time for stories and the jocularity of the saloon. Marcus had an affinity for the mining work, and there was much to learn. He was able to learn, earn, and send money to Ireland for his family. This was the land of opportunity, where real gold was to be found, along with a treasure of knowledge and opportunity.

Within a few years Marcus knew that his knowledge of mining and his natural ability for the work qualified him as an expert. The next step was to go to Virginia City, Nevada. A big gold and silver strike was there, and there was a chance to stake his own claim. Thomas decided to go to Virginia City with him. Thomas had cast his lot with Marcus, and thus far their fortune had been good. The town turned out for a farewell at the bars. The mine manager was included, for he was sorry to lose his best worker. Actually, Marcus was the only one of the men that the manager could converse with about prospecting, the geological formations that contained ores, working conditions for the miners and share a beer and some stories as well.

In the spring of 1865 Marcus and Thomas left Placerville and rode on horseback to Virginia City. The country was lush and green as they rode up to camp for the night. Lake Tahoe was like a jewel, reflecting the sunset in its waters. It was virgin territory, and they took precautions to stash their food sacks well away from their sleeping area. The bears were plentiful, as were fish, birds, wolves, mountain lions and other predators. Some of the predators were human, but Marcus felt their out-of-the-way camp for the night was safe.

The lack of law in the West required Marcus to carry a gun. The fire at the camp was left burning all night, and though Marcus and Thomas slept with one eye open and their guns at hand, it was a peaceful night. With the sun up, Marcus went down to the lake with a pole, string, hook and earthworm, to catch a small lake trout for his and Thomas' early morning feast.

Needing to get a good ride in before sundown, the men packed their gear, cleared up their camp and covered their fire hole with dirt so that one would be hard pressed to know there had been campers at that spot. They rode briskly through the mountains and the forest to Carson City, Nevada. On leaving Lake Tahoe, they began the descent from the Sierra Nevada mountain range into Carson City arriving early, for it was just twenty miles from Lake Tahoe. Carson City -- named for Kit Carson, the fur trapper and explorer -- was platted in 1858, and was established as the county and territorial seat in 1861. The ride Marcus and Thomas took was the Pony Express route from Sacramento, California, to Carson City. The dust-filled streets were teeming with the usual horde of lumber wagons and miners buying supplies for prospecting; and the boardwalks were lined with feed stores, a food store, saloons, and even a few rough private houses. It was the epitome of a boom town. Within twenty miles of Carson City was Virginia City, the famous Comstock Silver Mine and other strikes in the area such as Gold Hill, Aurora, Austin, Eureka, Hamilton and Belmont. Each strike brought with it a tent city, but Carson City was the trade center. Marcus and Thomas spent the night here, catching up on the news and listening to stories of the rags-to-riches mine strikes. The gold and silver mines here had made, for their lucky owners, a number of impressive fortunes.

After dinner Thomas and Marcus went to the boarding house to sleep, planning an early start for the next morning. The horses would need a good rest and a good measure of oats to be fresh for the journey to Virginia City. Just as they were ready to retire, they heard a paper boy shouting, "Extra! Extra! The Civil War has ended! The Union has won the war!" Hearing all of the shouts and hurrahs outside, they decided to go back to the saloon to celebrate the end of that great war, a tragic war of brother against brother, with its terrible destruction of the South. Marcus reflected on how strange it was that he had been on his way to California when he heard the war had been declared in 1861, and here came the end just as he left California. Carson City was in Nevada territory and they had sent wagonloads of gold to the Union forces for their war chest, which had greatly helped the Northern cause and President Lincoln. This, in turn, prompted President Lincoln to draft an agreement that led to Nevada's admission to the Union. There was much back-slapping and many rowdy toasts to the end of the strife.

Talk turned to Virginia City as Thomas and Marcus told their new-found friends their plans. The fortunes were being made by John Mackay, James Flood, James Fair, and William O'Brien. Timber from the Lake Tahoe area was used to shore up the mines. Marcus was eager for morning when he and Thomas would be on their way to seek their fortunes in Virginia City, Nevada.

After the hullabaloo and hurrahing of the Carson City Civil War Armistice celebration, Marcus was glad of the relative quiet of the stage coach headed for Virginia City. Only two other passengers were aboard; both seemed to be traveling salesmen somewhat worse for the wear from strong drinks and the strong feelings associated with the culmination of the long bloody conflict. They dozed, and Marcus gazed out at the dusty hills, thinking and wondering how long it would take for the nation to heal its terrible wounds. Even in the West, far away from the actual battles, vocal pro-Southern sentiments had been heard everywhere, especially from newly arrived settlers who had fled the prewar strife in Missouri and Kansas. Though the mineral riches of the West had enriched the war chest of the Federal troops, and most people were in sympathy with the Northern cause, there was division in many of the towns, and hard feelings abounded. Marcus sighed and thought, "It'll take at least a generation before we can put this war behind us. Maybe more." What a burden this had been for President Lincoln, who was faced with having to guide the country to wholeness again. Marcus felt a little twinge of guilt that he had not tried to volunteer; but even if he had, his mining was considered a necessary industry, so it was probably better that he served the nation as a miner.

Virginia City, Nevada was not just another mining town. It had some substantial and elaborate houses built by the newly rich mine owners. The John Mackay house was a show place, and as the money poured in from the silver and gold, other mansions were built. As people acquired huge amounts of wealth, they left behind the towns of Virginia City and Carson City for New York and Europe. They kept homes in Virginia City, but wished to live in luxury with the cultural advantages and comforts that only a large city and the society of New York, San Francisco and Europe could offer. San Francisco held an attraction for these mine owners, and many lived there. Much of the financial backing for the operation of the mines came from lawyers and bankers from San Francisco. Within a short while after arriving in town, Marcus and Thomas were able to get work in the Comstock mine, and Marcus did some prospecting on his own. He would find likely areas and then try to interest prospectors in the property. For this, he would either charge a fee or keep a small percentage of the claim. Meanwhile, Marcus became acquainted with a prospector named George Hearst. They spoke the same language and became good friends. George liked Marcus well enough to invite him to meet his wife. Marcus spent some enjoyable evenings with his new friends.

During one of his evenings with the Hearsts, George told Marcus he had monied partners in San Francisco who backed his mining operations. Marcus found this very interesting. Now he knew that there was hope for him to be on his own someday. Of course, he would have to find backers with money to invest, but with his mining expertise it was possible.

The next summer after Thomas and Marcus arrived in Virginia City, George Hearst asked Marcus to look after his mining interests, for he was going to check on the Alaska gold field that had been discovered. Marcus was glad to do this for his friend, and since Marcus had boundless energy he found it no burden at all. By this time Marcus and Thomas had their own claims and were working them. The mine was making enough to keep them going, but not enough to make them rich. Still, working the mine further educated Marcus in the ways of mining and geology.

The following May, Mrs. Hearst became concerned when George had not returned from Alaska, and by May she asked Marcus if he would try to find George. She offered him a fair wage and expense money for the trip. Marcus left Thomas in charge of both his and George's claims in Virginia City.

Following the Lewis and Clark Trail that leads over the Bitterroot Mountains into Montana through Idaho, Marcus traced the trail that George Hearst had planned to use on his return. After several days of horseback travel, Marcus came to Salmon, Idaho, where Lewis and Clark found the Salmon River that they thought would lead them to the Pacific Northwest passage that they were seeking for the U.S. government. Unfortunately for them, there was no passage to the Pacific Northwest here, for the Salmon River ended by going underground for some miles.

From Salmon, Marcus entered the Bitterroot Mountains and climbed to the summit. Below lay the Bitterroot River coiled in its beautiful valley. The sight took Marcus' breath away, for he had not seen such "green of grass" since Ireland. The sun glittered on the bunch grass and the graceful cottonwood trees, and the palate of greens was iridescent. The valley had the massive Bitterroot range with its many canyons on one side of the valley and on the other side were the Sapphire mountains -- a more gentle range, not so foreboding as the Bitterroots. Marcus felt a rush of homesickness. Not since Ireland had he seen a land to compare to the Irish countryside in beauty. Traveling on into the valley, Marcus vowed to return to this valley. Perhaps it was a valley where he could retire someday. He felt a great sense of being at home there.

While still traveling and daydreaming, Marcus caught sight of a group of men in the distance. As they approached, Marcus realized it was George, and the two men simultaneously spurred their horses. There was much rejoicing as George and his companions were found well and safe.

They camped close to the stage station at Victor and exchanged stories. George was not tempted by the strikes in the Klondike, for he intended to stay in Virginia City. It was a grueling trip to get to Alaska, a real killer of men and animals. No, George could do all the prospecting he needed in Nevada. Besides, Alaska gold was not yet proven. The next day, they climbed back over the steep pass and into Salmon. Then, they rode on into Utah and Nevada. It was not an easy trip, but stories of the Alaskan wilderness, the polar bears, the icebergs, the snow pack, the hardship, the winter's midnight sun kept the time from dragging.

As they got close to Virginia City, the horses and men both knew home was near and the horses' pace picked up. Their spirits rose, and when they arrived at George Hearst's home, a glad shout went up from Mrs. Hearst, "Come, George is home!" Mrs. Hearst burst out of the house, ran to George and practically pulled him off his horse. Even in those days of sedateness and careful manners, joy overtook manners. The men traveled to their homes, and Marcus rode to his rooming house, knowing that he left his friend in good hands.

Not too long after George Hearst returned from the Klondike, a son was born to George and his wife. The baby, William Randolph Hearst, he went on to become as much a child of destiny as Marcus was.

The Walker Brothers from Salt Lake owned the Comstock Silver Mine, and they heard about Marcus Daly, his ability to assess mining property and his reputation as a hard worker. With these capabilities in mind, they offered Marcus the position of mine manager at their Ophir, Utah, mine. They also asked him to do some prospecting for them. Marcus discussed this with his friend Thomas Murray. Thomas told Marcus he should accept their offer. The two men had been friends and partners, but taking the job in Ophir meant that Marcus would go alone. Marcus wanted Thomas to come with him, but Thomas chose to stay in Virginia City and work the small claim. Marcus made Thomas promise that when the mine played out or if he wanted a change, he would come work for him. Marcus accepted the Walker brothers' offer and made ready for another move. It was both a geographic move and a move toward a successful career and his destiny.

Marcus and Thomas shook hands and clasped each others shoulders and had their short goodbyes. On the morning of April 12, 1870, Marcus left for Ophir, Utah, by stage. As he boarded the stage and thought of the date he realized that he had made a move every five years since arriving in America.

Chapter Two

When Marcus arrived in Ophir it was mid-afternoon, he went immediately to the mine office to present himself. The mine personnel were waiting for him, and he was introduced to his miners and the office staff. The people seemed happy to have him, and he felt their cooperative attitude would help him immensely with his first trial as a manager, not a miner. After the introductions, the office manager, Mr. Willis, took Marcus to the mine manager's house.

They passed a row of five wood houses and a large bunkhouse for unmarried miners with its adjoining wash house. There was a cookhouse with a large dining room, a pantry and a black wood-burning stove for cooking.

The manager's house a white wooden bungalow with a kitchen, a living/dining area and two bedrooms, was comfortable and larger than the others in the camp. The house had sturdy oak furniture with dark, mottled upholstery. Certainly there was not a cheerful room in the house, but it was practical for everyday use. This suited Marcus just fine. He didn't need many creature comforts. Marcus cared little for possessions. It was the work, the study of ores and the prospects of mines that excited him. Leisure meant little to Marcus. It was work and results that gave him satisfaction. Marcus' recreation was to have a pint of beer with his men in the mines or with the mine owners and other mine managers.

Shortly after he became operator of the Ophir mine, a Mr. Zenas Evans and his daughter came to the mine. Mr. Evans was a lumber man and a prospector, and his daughter, unlike most other women her age, was interested in business. The two came together to inspect the mine, to determine the amount of lumber needed to shore the mine and to submit a bid for the operation.

Marcus was introduced to Miss Evans by her father, and after a short conversation they proceeded from the office to the mine. It was not far from the office door to the mine entrance and the three walked alongside each other. Mr. Evans was in the middle of the threesome and Marcus was forced to lean around Mr. Evans to speak to Miss Margaret Evans.

"Miss Evans, I'm delighted to have you on our mine inspection and will have to admit I have never had a woman enter the mine for business reasons," said Marcus.

"I know it is unusual, but Father needs someone to work up the bids for lumber, and I like the challenge and the travel the work requires. I assure you that I am capable of helping Father make bids, aren't I, Father?"

"Marcus, you can't imagine how much easier my work has been since my daughter has taken over the figures for the lumber needed."

As they entered the mine, Marcus cautioned Mr. Evans and his daughter to watch their footing as they descended into the mine. It was steep all the way, but even more so towards the end. Marcus entered into the mine first, Margaret came next, and Mr. Evans behind Margaret. Near the bottom of the mine Margaret started going faster and was about to tumble when Marcus, hearing her gasp as she lost control of her footing, caught her. When he caught her she was trembling, breathless, and grateful to be saved from an ugly fall. It was like a scene from a romantic novel. Marcus, who had not had much time for women, felt as if he had been stabbed. This gorgeous creature was caught in his arms, and he hoped that this intense sensation he felt wouldn't strike him dumb or make him lose his footing.

Margaret felt foolish, thinking, "Lost your footing? Mr. Daly probably thinks I did it on purpose -- how humiliating." But she had felt his great strength as he caught her, and she had to admit that he was the most dynamic, strong and courteous man and that she liked being in his arms. There were apologies from Margaret and reassurances from Marcus, and with her father's concern, the awkward occasion was righted. Finding no injuries, the three continued on the mine tour, noting the amounts of lumber needed to shore the mine.

Margaret and her father planned to figure the job and present a bid to Marcus the following day. Generally after Mr. Evans submitted his bid he would invite the mine manager and wife to dinner at the restaurant in the hotel. Since Marcus wasn't married, it would be just the three of them for dinner. Mr. Evans submitted the bid in the early morning and issued the dinner invitation to Marcus later in the day, after the Evans' lumber bid had been selected. Marcus accepted the dinner invitation, wondering what his reaction would be at seeing Margaret again. He couldn't forget his closeness to her in the mine.

"Could it be fate?" he wondered. He had never had this sensation before. Could she be his love for life? The idea sent a warm sensation all through Marcus' mind and body. He shook his head, for these new feelings made him wary. Besides, Miss Evans might not feel the same about him. He would try to convince her to care for him by example. Marcus thought to himself, "I'm honest in my dealings and loyal to my friends and workers. I'll care for Margaret with the same integrity." These thoughts shocked Marcus.

He had not allowed anyone to touch his soul, then this young woman had staggered him as if with a physical blow. Having never felt this way before, he had to wonder if this was what the word "love" meant, this pain, anxiety, physical weakness and then euphoria. As Marcus readied himself for dinner he looked in the mirror and wondered if his black hair turning grey on the edges made him look too old. Was his suit too dull, was he too short at five foot ten? He knew he wasn't too heavy -- working in the mines kept him physically fit and strong. Marcus hoped that he would measure up to Miss Margaret's idea of a beau.

Margaret was fretting as much as Marcus was. She was so embarrassed to think she might have appeared brazen. Also, the memory of being so close to Marcus Daly took her breath away, and if she had had on her tight corset she might have needed to recline to still the vapors. Margaret wondered what Marcus Daly thought.

Did he think her long brown hair caught in a bow was becoming? Did he like her size -- small, thin and strong-bodied? Would he like her blue dress that matched her blue eyes? Would he like her? Margaret thought it might have been better that they never met for she felt so unnerved by the thought of seeing him again. There was a knock at her door, and Margaret's father called for her to come meet Mr. Daly in the lobby and have dinner.

Both Margaret and Marcus were very controlled when they met for dinner. Mr. Evans carried on a lively conversation with Marcus about mining, prospecting and other personalities in the business. Margaret also contributed to the business conversation, but both were aware of the other being close. Through the following weeks Marcus and Margaret fell in love.

Mr. and Mrs. Evans liked Marcus Daly, but they thought their eighteen-year-old daughter was too young to have fallen in love. Nevertheless, anyone could tell immediately that Marcus and Margaret were committed, so the Evans decided that, even though Marcus was twelve years older, he would be a good husband.

Margaret had associated with miners, businessmen and their wives, as well as with young men and women her own age. She had attended high school, but her work with her father took her knowledge far beyond that level. Her social graces were more that adequate, and she would make a fine mine manager's wife. Marcus courted Margaret at her family's home. They spent every spare moment getting to know each other. They became good friends as well as lovers.

When Marcus proposed to Margaret he had already let his intentions be known to Mr. and Mrs. Evans, but Margaret seemed unaware that Marcus would propose marriage. Actually, she was so full of thoughts of her next meeting with Marcus that she never thought of what the more distant future might bring.

The actual proposal brought Margaret to tears. She reached for Marcus' hand and held it, telling him that she was overwhelmed with the thought that their love might be consummated in marriage. She finally said, "Yes, yes, yes! More than anything, I desire to be your wife and to spend my life beside you."

Marcus and Margaret went into the dining room to tell Mr. and Mrs. Evans of their commitment to one another and of their desire to be married soon. From there, the news traveled through the mining community, and everyone was happy for them. It had been obvious to most of their acquaintances that the two were in love, so the engagement was not very surprising. The Walker brothers in Salt Lake told Marcus they wanted the wedding at Joseph Walker's home in Salt Lake City, where an Episcopal minister would perform the ceremony. It was to be a ceremony that would join Marcus and Margaret as one to start their new life together. In the spring of 1871 the wedding took place in the elaborate house filled with the aroma of the lilacs that overflowed the baskets on the altar. The lattice work altar was erected at the end of the ballroom in a sea of tulips.

The dressmaker had employed all of her skill in making the dress for Margaret's lovely slim figure. Margaret wore a white wedding gown of lace over white silk and appliqued with white lace flowers with pearl centers. The neck was high and the tight bodice accentuated her small waist. She wore a gold circle broach with diamonds at the neck, and carried a bouquet of white roses with silk streamers and bows tied at the ends. Her headdress of orange blossoms and lace cascaded down the back of the dress into a small train. All the creativity of the seamstress was made more beautiful by Margaret herself.

The reception followed, and as the happy couple received the guests, silver trays laden with baron of beef, whole turkeys, fresh trout, breads, spiced fruits, and cheeses were spread on the tables on either side of the room. The cake, three tiers tall, had fresh lilacs around its edges and pink roses strewn among lilies of the valley. The champagne was French, and it was plain that the Walker brothers wanted to reward their mine manager and his bride with a ceremony they would always remember, and with the finest of food, wine, and friendship.

After a short honeymoon in Salt Lake the newly married couple returned to Ophir and the mine manager's house. The dismal house in Ophir was furnished in true bachelor taste. The living room furniture was in a dilapidated state, and a woman's hand and some carefully spent money could do nothing but improve it. Margaret was a frugal person and she knew a bucket of paint might produce a miracle.

Marcus told Margaret to have her way with the household that day and always.

Margaret's dowry added to the decor and warmth of their home. The walls were painted a soft green, the woodwork was varnished, the chairs were recovered in emerald green horsehair fabric, and the brass lamps were polished. The china was put in the breakfront in the dining room, with an oriental rug under the table. Again, Margaret used emerald green for cushions on the chairs. Marcus' Irish green always would be in their house. He never suggested green to Margaret, but she had observed that he always liked anything green and often mentioned the "green of Ireland". The redecorating was successful and Marcus approved. Most of all, Marcus loved to see Margaret in the setting of their home. Her warm smile, her cheerfulness and her eyes that lighted when she saw him made Marcus almost burst with love and gratitude. They were at home together.

In August of 1872 a daughter, Margaret, was born to them. She was a beautiful little girl with light brown fuzz for hair and very blue eyes. The pregnancy and birth were not unusually difficult, but no birth is easy and Marcus had hired a woman to stay to help Margaret with baby Margaret for at least three weeks, night and day.

Before baby Margaret, Mrs. Daly had given teas for the wives in the camp. Sometimes there were dinners for visitors or friends, but for three months this entertaining was halted so that the Daly family could become a unit.

After three months, Marcus returned to Virginia City for three weeks to look at a mine for the Walker Brothers. He looked up his friend George Hearst, and they caught up on each other's lives. The claims Hearst had were doing all right, but they'd produced nothing spectacular. Marcus told Hearst he was assessing properties for the Walker brothers. On his own, Marcus went to a mine that he thought had potential for silver production. After looking the claim over, he added this mine's assessment to the other two he had been assigned to inspect for the Walker brothers. While Marcus was still in Virginia City, the Walker brothers sent word to him that the option on the mine he recommended was not going to be exercised. Marcus was disappointed. He wished that he had the money to buy it himself, because he knew in his bones that it contained a big payout in silver.

On his last evening in Virginia City, Marcus and George met for drinks at the saloon. Marcus told George he ought to talk to his monied backers, Lloyd Tevis and Ben Ali Haggin, about financing the purchase of the mine the Walker Brothers discarded. It later turned out to be the first strike of any magnitude that George Hearst and his partners ever had. It proved to be worth millions. Without any formal schooling, Marcus could geologize a site, enter a mine, examine the veins and accurately assay the worth. The investors from San Francisco, Tevis and Haggin, were Kentucky lawyers who had not known each other before becoming business associates in San Francisco. Though their mining investments lay solely with George Hearst, they also had many sucessful investments in real estate, railroads and banks.

Returning to Ophir, Marcus found Margaret looking thin and wan. Baby Margaret was thriving but required frequent feedings and care that sapped Margaret's strength. As Marcus embraced her and felt how prominent her bones were, he decided that someone was needed to help in the household. He felt guilty for having left her alone for a month so soon after the baby's birth. He vowed never again to neglect Margaret's delicate health — not if he could help it and if God would help him. Margaret told Marcus she was fine, and that all new mothers were thin, especially with the first child. Margaret was the youngest of her brothers and sisters and she had not had much experience with babies. Every cry of baby Margaret sent a fear through Margaret that something was wrong. Actually, her maternal instinct served her well and she proved to be an excellent mother, forming strong bonds with all of her children.

Marcus hired Ann Tobin, a miner's wife who had no children of her own and who was happy to find work that involved a baby and the mine manager's family. Ann felt she was a special person to be allowed to serve in the Daly household. Though she had not realized it, Margaret found that she did indeed need help. She felt much better in only a week's time. Marcus was attentive when he was home, but he worked from dawn until a late hour every night. Dinner was served at six o'clock and Marcus usually arrived on time, but he would go out again at 8 o'clock to be sure all was well at the mine.

Marcus remembered clearly his days as a miner. He would always respect the men who worked for him and their contributions to the mining world. Because of his caring, he worked to be sure that the mines and the working conditions there were the best possible. Marcus also was a leader to his men and, like a Spanish *patrone*, he helped them in their crises of health, family and well-being. Marcus' theory was that a man should be able to do everything he asks of others. His philosophy was to respect all persons and their contributions to a successful operation. As a result, the workers' morale was always high in mines that Marcus Daly managed or owned.

The Daly family prospered. In the household, Ann's care for Margaret Daly and her baby made a difference in the health and happiness of both mother and child which in turn helped Marcus. Marcus looked in wonderment at his family. As an Irishman, he carefully noted his good fortune and knew he must not become too elated. The Irish believe that misfortune is not too far behind if one is too exuberant about one's luck.

Although Marcus was away from home many days at a time, Margaret was happy as the manager's wife and she kept her spirits high. Marcus let Margaret know that she was always first in his life. She knew that even when he was away from her, she was in his thoughts, and his hard work was for her. Marcus was able to convey his devotion to Margaret, and though she might be lonely, she never would feel unloved.

When Marcus returned from another trip to Virginia City, he found that he was to be a father again. Baby Margaret would be over two years old when the new baby was born. Marcus instructed Ann to be sure Margaret ate well and rested, for the new child was wanted. Marcus felt blessed, but he also worried that Margaret might become frail. She did get tired, but she had not had an illness. Baby Margaret was a quiet, sweet baby whom Margaret adored and played with for hours.

When discussing the new baby after dinner one evening, Margaret asked Marcus if perhaps they should start calling baby Margaret by another name. She no longer would be the baby, and two Margarets in one household was confusing. She would be "Madge". It suited her and, now that she was two, she was a little lady and no longer the baby Margaret. That same evening Marcus had to tell Margaret that the Walker Brothers wanted him to inspect a mine in Walkerville, Montana. It meant that he would be gone for a considerable period of time while she was carrying the new baby. It was poor timing to leave Margaret with so much responsibility and in precarious health. When Marcus told Margaret that he must leave her on her own but that, God willing, he would be back for the birth, Margaret was so quiet that Marcus was afraid she was angry. She was not. Margaret's mind was swirling with the thoughts of being alone and of Marcus being on a journey that could be dangerous. What if she never saw him again? At last she gained her strength and told Marcus it was not the best of news, but that she knew when she married him that her life would have a loneliness and responsibility that many women never experience -- especially not while expecting a second child at the age of twenty-two.

It 1875 the Walker Brothers assigned Marcus Daly to Walkerville, Montana, to inspect a silver mine called the "Alice".

From Ophir, Utah, to Salmon, Idaho, with stage stops on the Lewis and Clark Trail across the Bitterroot Mountains into the Bitterroot Valley, was the same route that Marcus had taken when searching for George Hearst. At the summit of the Bitterroots, the valley could be seen 4,000 feet below for forty miles. Again Marcus was struck by the grandeur of the Bitterroots. The lowland was green, the mountains blue and thick with pine, and the sky went on forever.

The stage rocked and the passengers clung to its sides as the coach and horses plunged down into the valley. The route, steep with twisting switchbacks, was a dangerous ride for horses, men and coach. As the coach came down the mountain, the drivers used the brake to keep the coach from running the horses down with a resulting catastrophe. At the bottom of the pass the stage road followed the river twisting its way through the wide canyon. The canyon extended into a broader valley, and the majestic Bitterroot range covered in pines stretched along the entire west side of the valley. Canyons with wide sides rose up into peaks and when one looked into the canyons a vast rugged wilderness with jagged mountains was reflected back in a breathtaking sight. On the east side of the valley were the gentle Sapphires, named for color of the mountains and the sapphire mines found in them. Between the two mountain ranges were the luxurious green grasses of the valley. Marcus experienced a "soft day" as he had in Ireland. The soft mist touched the face so gently with a balminess not felt in the Rocky Mountains areas of Nevada and Utah.

The stage stop was at Victor, Montana, and they reached it none too soon. Everyone was tired, hungry and in need of liquid refreshments and a liquid bath with soap. They were happy to have a bed and a good meal. There was no fellowship that night. It was an early evening for all, and the stage left again at 7 a.m. for Missoula.

The valley had long days in the summer. The sun rose at 5:30 a.m. and set at 10 p.m. It was good to have daylight because the way was not void of dangers from robbers and Native Americans. This stage did not haul gold or silver, so everyone was hopeful that they would not be a target. They had to go through Hell Gate, a pass on the east side of Missoula. It was narrow, and a stage coach or a wagon train or people could be trapped in this passage and robbed or killed by marauding Native Americans or professional robbers. Because of the "hell" getting in and out of Missoula safely, the pass was named "Hell Gate" by Missoulians. Marcus knew all about Hell Gate and was prepared with his pistol. Both of the stage hands had extra shotguns. Perhaps this stage would not be attacked, as it didn't seem to carry anything of value.

Five valleys came into Missoula like spokes on a wagon wheel with Missoula as its hub. Missoula is on the Clark Fork River, which has the Bitterroot River enter into it at Missoula. The Flathead River empties into the Clark Fork west of Missoula. The Clark Fork River then carries its waters to the Columbia River in Washington and on to the Pacific Ocean.

The town was a hub for trading. Native Americans and trappers engaged in trading furs, cattle and horses. It also bustled with miners and locals getting their equipment, food and supplies. The mountain men, miners, travelers and merchants came together for their common interests in Missoula.

Fort Missoula had a full complement of soldiers, and beyond its walls were the traders and Native Americans. At night one could hear the music, shouting and laughing of the fur traders, horse traders and now and then a Native American call from the tepee village. The noise didn't bother Marcus. He was exhausted and sleep came quickly.

The stage left in the morning and the group made it through Hell Gate without an incident. Marcus felt prepared to handle himself in an emergency. He certainly had too much to live for to die in a robbery. He kept a watch along with the outriders that escorted them through the dangerous area.

This assignment to assess the Alice mine would be a problem, for it was a "tight" mine. No information or inspection was allowed or could be expected. Since Marcus had been a miner he would hire on as one. He finalized his plan on the third day of the journey to Walkerville. Upon his arrival he found a boarding house and asked the owner, Mr. George, to help him get a job at the Alice mine. Mr. George needed a boarder and thought Marcus looked strong and healthy and seemed to be an experienced worker. Marcus was hired, allowing him to carry out his ingenious method for assessing mine property. Marcus worked three weeks, paid his board and room and prepared to return to Ophir.

The return trip took him back through the Bitterroot. He passed the Lolo Pass, which was the passage through to the north Pacific coast that Lewis and Clark had sought and found. Their Native American woman guide, Sacajawea, had told Lewis and Clark about this passage. The Clark expedition thought the Salmon River would lead them to the Pacific. Instead the river ended for a time by going underground. Sacajawea's tribe knew of the Lolo Passage and she remembered this.

On past Lolo was the settlement of Stevensville. St. Mary's at Stevensville was the earliest mission in the territory. The Salish Indians had requested the "black robes" start a mission at

Stevensville, and it was named St. Mary's. Father DeSmet was a priest familiar with the west and the Rockies, and he made his way into the valley, as well as the territories of Idaho and Wyoming. Father DeSmet established the mission, but the Indians became disillusioned when the Father wanted to welcome the Blackfeet Indians (their arch enemies) into the church.

Later Father Ravalli established a permanent parish at Stevensville. He was friendly with the Native Americans and started the first farms in the area. Marcus wondered how they survived for there was little water and few settlers to establish agriculture, life and religion at St. Mary's. Father Ravalli was the religious leader for the community and held daily Catholic Mass for the people in the valley and for travelers.

In the Bitterroot Valley, Marcus noted there were a few ranches, a small settlement in Victor around the stage stop, and some farms at Grantsdale. A group of ranchers had formed a co-op and had their livestock combined and fenced in what was called "The Big Corral". It encompassed about nine square miles near Grantsdale. All of these features, along with the Bitterroot River were filed for future reference by Marcus. It was a lush valley with abundant timber available and perhaps a place to live someday. Marcus liked what he saw.

Marcus had been gone two months by the time he returned to Ophir and Margaret and baby Madge. Marcus walked into the house, and Margaret turned at his call. Marcus was always blinded by Margaret's smile, especially when he returned from a trip. They embraced and confided of their love and loneliness at being apart.

Marcus noticed a desperation in Margaret's embrace and felt tears on her face as they clung to each other. Marcus knew how hard it was for her to be alone — harder for her than him, for he was leaving her in a safe place and he was courting danger whenever he left the house. Understanding this, Marcus reassured Margaret that he was ever watchful and careful, for his love and life was with her. At that, Marcus and Margaret went into the dining room where Madge was having her dinner. There was porridge on her face and bib, nevertheless, Marcus lifted her out of the chair and held her close. Madge squealed with happiness to be in her daddy's arms. Madge always said, when Marcus left for work, "Can I go 'witch you, Daddy?" Madge was approaching three years and busy all day, but she did drop off to sleep after her dinner at six o'clock.

Marcus reported to the Walker Brothers that the Alice mine was worth the price and should bring a handsome profit to them. Part of Marcus's pay from the Walker Brothers was a 5% interest in the Alice mine if and when they purchased it.

Marcus visited another mine for sale in Butte. Two fellow Irishmen wanted to sell their silver mine, and Marcus was curious about the sale. The silver vein did not run too deep but, interestingly, there was a large copper deposit in the mine. Marcus felt that it was a copper vein that would be more valuable than the silver and would last much longer. When Marcus offered the Walker brothers this mine, they said they would think about it and wanted further evaluation. They told Marcus if they purchased the Alice mine he was to manage it. This meant a move, and Margaret was going to have a baby. The earliest she could move would be in the spring of 1876, when the baby would be old enough to travel the rough stage passage to Butte.

The Walker Brothers did buy the Alice mine and Marcus

was to be the manager and owner of a 5% share. At the same time he was to manage the Ophir mine and plan the operation of the Alice mine in Walkerville.

Margaret went into labor on the night of January 4, 1876, and again the birth of a baby girl was not a difficult birth. Margaret was young, and her self discipline and good health helped her have an easy birth. There was never any worry that Marcus would be disappointed that another girl was born. He assured Margaret that girls were fine with him, and if they had a boy that would be all right, too. Marcus was relieved when the birth was over. What Margaret or others thought was an "easy time" was not so in his mind. He worried that Margaret might be injured, but his worries were unfounded, and Margaret greeted Marcus with her wonderful smile and a blonde baby girl. The name chosen for this daughter was Mary Elizabeth Daly. All rejoiced at the birth in the Daly household. Friends and workers all called to see the baby and bring gifts to welcome the newborn.

As soon as it was spring, the Daly family would move to Walkerville, Montana. This would not be an easy move with an infant and a two-and-half-year-old. Only treasured personal items would be taken to their new home.

The spring came early that year, which made preparations and the journey easier. The wagon with their belongings would be brought to Walkerville as freight, and the family would travel by stage. A shorter route to Walkerville from Utah was in use now, and it took the stage across the flats of Utah to the higher elevations and through a small pass into Butte. They rode through great plains of grass with subterranean water surfacing in streams, and lush banks holding the lazy flowing water that wended its way through the valley called the Big Hole. Buffalo, deer, moose and elk came to drink at dusk at the water holes. This was a beautiful area and a pleasant journey. This route was not as dangerous as the sharp descent into the Bitterroot Valley that Marcus had taken twice before. At best it was still an arduous journey, but Margaret was able to cope with any crisis that would come in the Daly family. As long as Marcus was with Margaret she could endure the hardships. Margaret was secure in her love for Marcus and in his ability to take care of her and the children. Since this was their relationship the journey was enjoyable because they were together. The family was off on a new adventure and both Marcus and Margaret knew that whatever would be their destiny it would be together.

Walkerville was like other mining towns, with streets that were dusty in dry weather, and ankle-deep in mud when wet. On its arrival in Walkerville, the stage stopped at the hotel, which was the stage stop as well as the town's center of activities. While Marcus searched for transportation to their house, Margaret and the girls waited in the hotel lobby. It was not long before Marcus returned with the carriage, and the Daly family proceeded to their new home in Walkerville.

The town was support for the workers, and stability and comfort were not a high priority. Wooden-sided tents for the miners (called tent cities), a general store, saloons and the hotel made up the main street of Walkerville. The mine manager's house was a wooden structure much like the one in Ophir, and the Daly family settled into their new house. In no time at all, Margaret turned it into a home. She had a natural flair for making a room warm and inviting, and for doing so with a minimum of money and materials.

Again, she used the rich green as the basic color and used gold trim for warmth and richness. The company furniture was transformed from the ordinary into the outstanding.

The family was settled, Marcus was having success with the Alice mine and their lives were quiet at last. Marcus came home for dinner each night. There were no more long journeys in sight, and Margaret was grateful for their peace and happiness.

On June 26, 1876, one month after making the trek from Ophir, Utah, word was received in Walkerville that General Custer and over 200 of his men had been annihilated by the tribes of the Indian Nations who were joining together to confront the U.S. Army in a territorial war. The town was agog at the news of the fight at the Little Big Horn. General Custer had ridden into a hornet's nest of angry Sioux and Cheyenne braves instead of into the small village encampment described to him by his 7th Cavalry detachment, and for all practical purposes they had been annihilated. He was not well served by his adjutant officers Benteen and Reno, who became frightened and confused to the point of disregarding all of their orders as well as their training. The Indians had always been very ambivalent about George Custer, but then even his friends had mixed feelings about the eccentric officer. Though the Sioux didn't hate Custer as they did General Crook, who they called "Child Killer" they thought Custer was insane. In the names they gave him, "Long Yellow Hair" and "Son of the Morning Star," there was a subtle suggestion that he was to be feared because he was not altogether rational. After the slaughter of the Little Big Horn fight, his was the only body not mutilated by the squaws. Like most aboriginal people, the Native Americans were in awe of anyone touched by the wings of madness. Custer's charge was routed but a number of them made their way to the top of the ridge. At that time, one thousand Cheyenne Indians under Chief Rain in the Face rode in, and not a trooper was left alive. All the bodies were mutilated except General Custer's.

The general population was aghast that Custer and his crack troops had been defeated, and wondered what the fate of commerce, farms and settlements would be. The U.S. Army answered their questions in short order, for the troops secured the area, and peace treaties began to be signed. It was relief for the white people in the west.

Chapter Three

The baby Mary became Molly. When Madge tried to say Mary, in her baby talk "Mo-rie," it sounded like Molly, and so Molly it was.

Although Marcus was not a heavy drinker, he did enjoy his "pint" with the men at the local saloon. On occasion he would be late for dinner. Margaret would greet him in a severe manner, and Marcus would react by being on the offensive. It was the only problem that the two had, and their communication was vocal and angry. While waiting for Marcus, Margaret would be apprehensive for fear he had been hurt at the mine, or that his horse had thrown him or simply because she didn't know where he was. Then anger would take over, and she would think Marcus thoughtless and wonder why he cared more for his male companionship and drink than he did for her and the children. At the sound of his steps on the porch, relief would surge through her; then her anger rose at the sight of his smiling face. Her fury and disappointment would last through the night, but Marcus forgot all about it as soon as he went to bed, and the sound of his soft snoring further infuriated Margaret. At last Margaret would fall asleep with tears still on her face.

After his evening with the boys, Marcus would waken with Margaret's stiff body beside him, and he would apologize for causing her such anguish and for being so callous to her feelings, vowing that it would not happen again. It did, and Margaret never could control her anger, fear and relief. Each time an episode happened there was the same reaction from both Marcus and Margaret. It was a cross for Margaret to bear, but Marcus thought his behavior was normal and could not understand Margaret's indignation.

The Alice mine was operating at top speed, and the silver gave a good monetary return for the investors, including Marcus. Because he received a top salary for his work as mine manager and prospector for the Walker brothers, plus the percentage of the Alice that he owned, the Daly family received a good sum of money, and life was comfortable.

Edward and Michael Hickey, brothers who owned the Butte Hill silver mine, were still trying to sell it. Marcus had told the Walker brothers about it before and did so again. Marcus was sure the copper vein that he had seen in the mine was deep and wide. This would make the mine viable even after the silver had been extracted. Again the Walker Brothers refused to buy the Butte Hill mine. Marcus was very disappointed and was unhappy that the Walker brothers would not take what he knew was good advice. Since Marcus had been hired to manage the Alice and to evaluate other mines as well, he felt insulted that the Walker brothers had rejected his recommendation. Marcus was angry, but reason prevailed and he decided not to pursue this any further with the Walker brothers. After some thought, Marcus had an idea that might appeal to the Walker brothers. He would sell his share of the Alice mine and buy the Butte Hill mine. He would then need investors for the operation of the mine. The Walker brothers might be one of the investors, so he would not alienate them.

Because Marcus was so confident the Butte Hill mine was a rich mine and therefore a good investment for everyone, he was astonished when Margaret was not receptive to his idea. Margaret, who would follow him to the limit and had never presumed to speak against his judgment, told Marcus that investing his life's work and money and his family's security in a mine that no one considered productive, was not her idea of a rational decision. At this advice from Margaret, Marcus said, "Well, my girl, you will go to the mine with me, and I will prove to you that I am rational. I believe in my assessment of the copper and there is still a good play in the silver, and it will be a very good investment." The next day, Marcus and Margaret went with the Hickey brothers to inspect the mine, and Margaret was convinced that Marcus was correct in his belief that the mine had tremendous potential. Probably the bond between Margaret and Marcus became stronger, for she agreed wholeheartedly with him and was willing to risk a stable, comfortable life for the roll of the dice on Marcus and his uncanny ability to gauge the value of a mine.

Joseph Walker would be in Walkerville the following week and that would be the time to talk business about the Alice and the Butte Hill mines. Marcus would approach the Walker brothers to sell his 5% of the Alice mine. With the 5% sum, he could purchase the Butte Hill mine. Believing in the mine's potential, Marcus would then ask them if they would like to be investors in the operation. This decision made, Marcus felt a weight lift from his soul, for he had been troubled by the rebuke from the Walker brothers and by Margaret's hesitancy about the mine venture. Now that Margaret agreed, the rest would fall into place.

Joseph Walker met with Marcus in the morning after his arrival on the stage Marcus was doing well with the operation of the Alice, but Joseph wished he would forget the Butte Hill mine. Joseph knew Marcus was irritated that they did not want the mine, but enough time had passed for Marcus to have cooled down. Joseph looked forward to seeing Marcus, gleaning the latest in mine news, and receiving a financial report of the Alice. The meeting went well until Marcus said he wanted to sell his 5% in the Alice so that he could buy the Butte Hill mine. Joseph was aghast. "Ye Gods, man, haven't we assayed the mine and told you it is not feasible for silver, and the possibility of a copper pay-off is not reasonable?" Marcus and Joseph argued, Marcus insisting that the copper vein was big and deep and that it would be a great play.

Joseph said, "Marcus, since you've given so much thought to this, I suppose you have a sum that you would take for your 5% of the Alice mine. We certainly don't want another investor in our mine. You do know that you will not manage the Alice mine anymore and there will be a replacement immediately." Marcus was taken aback, for he thought he might continue managing the Alice mine as well as starting up the Butte Hill mine. His salary from the Walkers would keep his family living well until the Butte Hill mine started paying off. Knowing that no matter what happened now, his relationship with the Walker brothers was finished, Marcus with no hesitation told Joseph the amount he wanted was $30,000. Joseph, irritated, said, "Sold!"

Certainly Marcus had mixed emotions. He was elated that the Butte Hill mine would be his and deflated that he no longer had a position as manager of the Alice mine, nor any hope of the Walker brothers as investors.

George Hearst was the first person who came to mind. Marcus would ask him to be an investor. George was his good friend and, since Marcus had recommended the Ontario mine, George and his investors had bought the mine and had operated it. The mine had been very lucrative, the first big strike George had had. Surely he and his investors would and could become partners in the Butte mining venture. Marcus immediately sat down and wrote George a letter stating his proposition and giving his facts and figures on the Butte Hill mine. Next Marcus vacated his Alice mine office, took his belongings to the buggy outside and went home to tell Margaret the good news and the bad.

When Marcus had left early that morning, Margaret's world was all intact, but now with Marcus' announcement it was shattered. Margaret pulled herself together and said to Marcus, "I'm happy you have the Butte Hill mine, but I'm concerned for you, too. This is a big undertaking and I'll do my part for you, Marcus. Tomorrow, I'll find a house for us in Butte while you finish your business with the Hickey brothers."

The Butte mine and its paltry silver deposit was about played out, but Marcus would keep a crew working, mining silver, which would support the men and his own family. The Hickey brothers quickly agreed to the sale, and on June 14, the deal was done. After signing the documents, there was a toast in the bar and much talk of mining and their common homeland, Ireland. Ireland was a troubled country, and the men wondered if it would ever be the glorious placed it should be.

Margaret had found a house in Butte, and the family moved in and settled quickly. George Hearst's letter came within two weeks. George was encouraging, but it would be a while before he could contact his backers, Tevis and Haggin, in San Francisco. This put Marcus in a bind because he was barely able to make ends meet and he hadn't found anyone to invest in his venture. Marcus decided to write Mr. Tevis and Mr. Haggin directly and send a note to George that he was doing so. They replied that they were sending their engineers to look at the mine. More precious time was passing, and finally the consulting engineers came to the mine. With great skepticism, they listened to Marcus explain how the vein was wide and deep because of this geological formation.

The engineer's report to Haggin and Tevis was a recommendation not to invest in this property, and a message was sent to Marcus that the engineer's report had been negative. Desperation was starting to cloud Marcus' judgment. He began to wonder if he was wrong and what his mistake might cost his family. Margaret decided that she needed to speak up and let Marcus know how she felt. It was hard for Margaret to see him so unhappy and frustrated. When Margaret had a chance to talk to Marcus, she told him that she believed he was right in his analysis of the mine and that he would eventually get backers to operate the mine, but that he should speak to them face to face. She suggested that Marcus go to San Francisco with drawings and charts to present the facts to Mr. Haggin and Mr. Tevis.

Going to San Francisco was a long and tedious trip by stage and railroad. As he neared San Francisco, Marcus thought how helpful it would be if the railroad crossed the United States. The railroad was planned and had been started, but lack of funds slowed the pace. The railroad was being constructed on the west coast to go east and on the east coast of the United States to the west, and would meet and join together somewhere in the middle of the country. It seemed a faraway dream to Marcus, but he certainly understood the impact a coast-to-coast railroad would have on commerce and transportation.

As Marcus neared San Francisco he made a list in his mind of reasons for Haggin, Tevis and Hearst to invest. He would finalize his arguments by offering a partnership in exchange for the financing of mine operations in Butte.

Hearing the conductor call out that the San Francisco station was the end of the line and everyone must vacate the train, Marcus quickly gathered his valise and portfolio and proceeded to the exit.

The air of San Francisco was so fresh and the temperature so stimulating that Marcus found new energy and hurried on to a horse and carriage that carried him and his belongings to a hotel. The city of San Francisco had changed since Marcus left in 1865, and he thought it was all for the better. The buildings were substantial, the port was busy, and foreign trade had made San Francisco a major commercial and financial center. Hearing a clanging noise behind him, Marcus turned to discover a cable car. He had heard they had them in San Francisco but had not seen them. What a great invention, Marcus thought, and a great help for the person who went from one part of the city to the other, for the hills of San Francisco were formidable for people walking or for horses pulling their loads. Marcus would ride one to the office of Ali Ben Haggin for their meeting the next day.

On the morning of that fateful day, Marcus decided to have his family breakfast of hot cereal, eggs, sausage, biscuits, sliced oranges and coffee. He needed his strength for the meeting with his backers, and this meal would surely do it. The cable car transported Marcus to his destination, and he stepped off at the front door of the office building and entered. In the lobby Marcus looked for the number of Mr. Haggin's office and then proceeded to the stairs up to the second floor. It was a very impressive building with green marble floors banded in black marble. The wood trim was rich mahogany and the staircase was massive with handrails and turned spindles. Marcus thought the green stair carpet was attractive, with its soft deep pile, and the stairs had proper risers which made the stairs easy to ascend. The office door had frosted glass with bands of mahogany forming rectangles in a pattern in the glass. Below the door was a brass plate that read Ali B. Haggin, Attorney at Law.

Taking a deep breath, Marcus opened the door and entered. The room was paneled and hung with heavy wine-colored draperies that cut out most of the natural light. Directly ahead was a male secretary sitting at a partner's desk. Marcus went to him and stated his business. At that, the secretary rapped at Mr. Haggin's door and announced Marcus Daly. Ali Ben Haggin rose from his desk and came forward to shake Marcus' hand. Although they knew of each other, the two had never met. They turned to the left into a conference room where George Hearst and Lloyd Tevis were waiting. Lloyd Tevis had not met Marcus, and proper introductions were made. Then it was time for old friends George Hearst and Marcus to greet each other, and they did so with warm handshakes.

Ali Ben Haggin was a quiet man with rugged features and dark hair. He was of medium height, standing eye-to-eye with Marcus. Haggin was a corporate lawyer and was consulted by many businessmen on matters of company formation and legal liabilities. Law was only a small part of Mr. Haggin's life, however. He was a shrewd investor and, with his partner Lloyd Tevis, he had achieved considerable success. One of their partnerships was with George Hearst, and that, too, had been a successful venture. Lloyd Tevis was tall and thin, with piercing blue eyes and fair hair worn unfashionably long. Tevis and Haggin were involved in a number of profitable enterprises ranging from irrigation projects in the arid California valleys to telegraph and railroad companies. Tevis was

President of the Wells Fargo Company, as well. George Hearst was first and foremost a prospector for silver and gold. His outdoor life showed in his tanned and leathered face. He was a handsome six-footer whose sharp brown eyes never missed a thing. George and his backers were successful in their mining operations. He was the founder of the Hearst fortune, and his son, William Randolph Hearst, founded the Hearst Publishing empire, the *San Francisco Examiner* and various magazines.

Marcus started his presentation with a short background speech. "Gentlemen, it never occurred to me that I would have any trouble finding financial backing for the Butte Mine I purchased. To me, the geological formations above ground as well as in the mine proper were so obvious that I was flabbergasted that the mine engineers sent to inspect the mine did not concur with my analysis.

"Though the mining engineers do not concur with my analysis of the Butte mine, I am not wrong about it and its copper potential. The silver will soon be completely extracted, but copper mining will take the place of the silver extraction. I believe that this mine will produce more copper than we could imagine."

Marcus unfolded his charts and placed them on the map board on the wall.

"Here on these charts you can see where the copper vein begins in the mine and continues upward to the formation above the ground. This formation above the ground goes for one hundred feet. Here is a core sample I took from below the mine entrance, which indicates the vein formation goes far below the mine.

"Gentlemen, I'm anxious for you to be partners in this venture. My hope is that we will form a partnership with each of us having one-fourth interest. I will be the President and

Chief Operating Officer, personally supervising the mining of the copper, the smelting and the distribution.

"There is a need for copper for telephone and telegraph wires, copper pipe, commercial construction and household use. Whatever the future brings to America, I tell you, men, copper is the king of the metals for use in an industrial nation, and we are becoming that!"

While Marcus was presenting his proposition, Ben Haggin was thinking of Marcus the man. Ben Haggin and others found it hard to believe that this immigrant Irishman was a natural in mining engineering, geology and management. Here was Marcus with his Irish brogue presenting a learned and detailed report that surpassed anything that had been presented to him by educated mining engineers. This Irishman, Marcus Daly had intense blue eyes that could look into a man or a mine and know what was there — a good or a bad person, a good or a worthless mine. Marcus was getting grey streaks in his hair and Ben knew most of the grey had come these last six months. He also knew that Margaret Daly had been taking in washing in Butte to help finance their household because there was no income from the mine. Marcus was without funds to operate the mine and was forced to close it. Ben knew Marcus was in great need of financing.

When Marcus finished his proposition, Ben Haggin said, "Marcus, I am with you. Here is my checkbook, but leave one dollar. When that is gone there will be more. If Lloyd and George do not care to join us, it will be their bad luck, I remember the largest strike George, Lloyd and I had was the Ontario mine in Lake Flat, Utah, that you recommended to George. I believe in you and your advice. If you say, Marcus, that there is a big copper deposit, I think you know what you're talking about. Besides, you've had great success in mine management. This will be a first-class operation." At this bold statement, Hearst and Tevis didn't hesitate to join in the partnership. Marcus and the three wrote out their partnership agreement. First a name had to be found. Marcus suggested "Anaconda" — the vein was huge in formation and its turns and coils reminded him of the South American Anaconda — a powerful, huge and long snake. Since there was no copper being mined as yet, the Silver Mining Company was formed and then, shortly after, The Anaconda Copper Company.

The return journey to Butte and to Margaret seemed longer than ever. Marcus was anxious to get home to his family. He hated that Margaret had had to help with expenses, but he would make it up to her. Marcus wondered at Margaret, for she pitched in and even told him she was pleased and proud to help so that the mine could be put into production. He promised himself that this would be the last time Margaret ever had to work at manual labor. He would make her life as comfortable as humanly possible. And, Marcus thought, "Was there ever a man so lucky as to have this beautiful woman as a wife, partner and help-mate?"

As the stage neared Butte it was getting dark and lights could be seen on the hills where mines were being operated. Between the sky full of stars and the lights dotting the landscape, it was as if a large city was ahead.

Marcus dashed to his home to his Margaret, Madge and Molly. What a homecoming it was, for he had been away for six weeks. Marcus told Margaret of the partnership, the money that was more than he had hoped for, and how thankful he was that Ali Ben Haggin had believed him. Marcus knew that without the endorsement of Haggin the other two would not have invested. The next day, Marcus told Margaret to have the laundry delivered and that there would be no more "take-ins" ever again.

The thought of starting the operation again on the Butte Hill mine (now the Anaconda) and exploring the copper vein would overwhelm most mortals, but not Marcus. He plunged into the reorganization of the mine, gathered his work force, worked side by side with the miners when necessary, contracted for lumber and got supplies. The mine was in production within two weeks.

Many of the workers in the mine had known of Marcus or had worked for him in the Alice mine. Marcus loved his miners. He never forgot how hard they worked. He saw that they had the best pay, the best housing, the best of everything. Marcus knew that without the miner the ore stayed in the ground. As a result, the men followed Marcus to any lengths, for they knew he considered them equals. The men felt Marcus was a benevolent dictator who had their interests, as well as his own, at heart. Marcus considered the miners and engineers his friends and treated them a such. So began the operation of the Anaconda mine. Marcus was ready to test the copper vein, and for the next five years more copper veins were unearthed that, like the Anaconda, snaked through the earth.

Michigan had the only other large copper deposit in the United States. The smelter there was not available to the Anaconda. The closest smelter for the ore from the Anaconda mine was Swansea, Wales. This was certainly unacceptable, and Marcus knew the company would need to find a smelter or have one built to handle their ore. The demand for copper for wire, pipe and telegraph lines was rising by the early 1880s, and the Anaconda mine prospered with the increasing demand. In Montana, the mining of gold, silver, copper, lead, coal etc. was providing wealth to some, employment to others, and investments to the wealthy on the East and West Coasts. With the building of the railroads from coast to coast and the good transportation for commerce, Montana citizens prospered. None of this was lost on Marcus, and he thanked the good Lord for putting him at the right time in the right place.

Now it was imperative to build a smelter, and Marcus chose a site 25 miles from Butte to build it. The smelter would need workers, and workers would need housing and supplies and, indeed, a town to support the smelter. The work began on the smelter and the town was platted. Houses were built for the management as well as the workers. Marcus knew the money saved by eliminating the Swansea, Wales, smelter would pay for the town of Anaconda before too many years. The smelter was ready by 1884.

The four men, Daly, Hearst, Tevis and Haggin received tremendous profit from the Anaconda mine and smelter. As a result all four became extremely wealthy. The Northern Pacific rail strecthed out from the east, meeting the workers from the West coast at Gold Creek, Montana, on August 22, 1883. This was a frantic time for Marcus as he master-minded all the progress and decisions for Anaconda Copper Company. He worked long hard hours and tried to be with his family as much as possible. It seemed to his associates that Marcus was super human to be able to carry such a work load.

The Daly family moved into a new house in Butte the same year Marcus II was born, a male heir to the Daly name and fortune. There was a glorious celebration by the Daly family and friends.

Margaret was lonely for her family and since Marcus knew

he would be unable to help Margaret or be with her because of the work, he asked Margaret's mother and father to move to Butte. The Evans family came, and it was wonderful. Margaret no longer felt lonely when Marcus had to give his time and energy to the mining company. Her mother and father and sisters (Mr. and Mrs. Zenas Evans and their daughters, Miriam, Hattie and Mattie) settled into their home in Butte. Their coming made home life and social life much happier for Margaret. Butte society found them a fine addition and accepted them just as they had Marcus and Margaret Daly.

When Marcus Daly and his family had moved to Butte, William Andrews Clark a leading citizen of Butte, accepted and introduced them. Clark, after years of being a supplier of goods, had gone into mining as well. W. A. Clark, Marcus Daly, S. T. Hauser and C. A. Broadwater became the Big Four in the Democratic Party of the Montana Territory. When W. A. Clark decided to run for the U. S. Senate seat representing the Montana Territory, the other three tried to dissuade him. They did not think he would be effective as a representative of Montana since the federal government was Republican, and the mining industries' indiscriminate use of federal lands for lumber had embroiled them in federal law suits. Since W. A. Clark was an owner of both mining and lumber companies, the others decided against Clark as a candidate and sponsored a young man untainted by the industries. This infuriated Clark. He turned his wrath on Daly, Hauser and Broadwater. Clark thought that Marcus in particular had betrayed him and was the power that had swayed the others, for Marcus, could deliver votes in any election from his loyal workers. Marcus tried to reason with Clark, but all was lost in their friendship. The feud that was to last 15 years became more bitter at each election, because Clark ran for office every time.

W. A. Clark had a well established newspaper where he vilified Marcus Daly's reputation regularly. Marcus was helpless against the printed word and later, in self defense, founded the *Anaconda Standard* newspaper. And so the battle raged. In one of his paper's articles, Clark said, "Haggin is a foreign Turk who tells Marcus Daly what to do and when." This caught Marcus off guard, and he decided then and there to get his own newspaper to defend himself and his friends and to attack Clark as a greedy, bigoted traitor. The feud built slowly, and life in Butte continued for the Daly family in an almost normal manner.

Prosperous Butte had an opera house and a busy social schedule that was not deterred by the political disputes and the feud growing between Daly and Clark. The Opera House had scheduled a company of singers from New York who came by train to Garrison. Garrison was west of Butte and served as a junction between the Northern Pacific and the Utah Northern railroads. The singers thought this trip was a marvelous adventure, and their performance was exceptionally well done. The Butte audience applauded so vigorously that more than the usual three curtain calls were given.

The women ordered their dress patterns from Paris; their fabric, braids and embroideries were of silks from the Far East; and their brocades and laces came from France and Belgium. Their dresses were as stylish as those in New York, which, of course, was the style center for the United States. Margaret, Mrs. Evans and Margaret's sisters were like pastel confections in their peach, blue, yellow, green and gold. The style of the period called for a tight bodice, a cinched waist and a bustle in the back of the dress. The bodices of silk were trimmed with lace and silk braid to match the color of their dresses. All wore their hair swept up with small curls at the nape of the neck and on the forehead. The family of women and escorts was impressive. At the opera house, they visited with friends in the lobby, then took their seats in the Daly box.

On the 12th of July, 1884, a lovely Daly baby girl, blonde and blue-eyed, was born. Her name was Harriot Holmes Daly, and Marcus called her his "little princess". When called "Princess," she responded with a smile.

Marcus tried to be home every night for the dinner hour with the children. He was vitally interested in their schooling and other activities. Of course, Harriot sat in the high chair and gurgled, and Marcus II was a normal two-and-a-half-year-old trying to assert himself. Margaret and Molly went to the Butte public school, but they also had tutors for French and piano. The household was filled with activity with Grandmother and Grandfather and aunts to make them all feel secure and loved. This way of life went on for Marcus and Margaret for the next four years and all was well except for the Daly-Clark feud that caused trouble for the children in school and for the Daly's friends. The town was divided.

Marcus felt at a disadvantage in telling his side of the story, for Clark used his newspaper to blast Marcus and Anaconda Copper Company every day. All of this was distressing to Margaret. Peace was something Margaret valued and life was anything but peaceful then.

Regardless, the mine and the smelter were facts, and the amount of copper being produced was making the men in the partnership millions. After its first few years of operation, the Anaconda mine was producing as much as its Michigan rivals, which were the only other large copper producers in the United States. With its new smelter, Anaconda had become very competitive in copper prices, too.

Margaret didn't want to trouble Marcus with the problems caused by this political feud, but the girls were unhappy and she was, too. Margaret decided that they would not have a peaceful household as long as they lived in Butte. Many of the people that had acquired wealth through mining had moved to the west or east coasts. Margaret believed more appropriate education and better cultural advantages for her children would be available in New York City. Now she had to find the correct time to approach Marcus. She knew Marcus well enough to wait until the right moment.

In 1885, Marcus and Margaret decided to move their home to Anaconda. Margaret had told Marcus that life was unbearable for the children and for her. She was sure the feud was hard on Marcus, as well. Margaret said, "Marcus, we should have a house in New York City." Marcus felt as if he had taken a physical blow, he said, "Margaret, you know I can't live in New York and take care of the business of Anaconda Copper Company. Perhaps we should move to Anaconda." Margaret was firm about New York, but consented to move to Anaconda and spend the school year in New York. She wanted a Fifth Avenue apartment, for she knew that was a respected address. Margaret wanted her children to have all the advantages that the Daly wealth could buy. That would mean summers could be spent in Anaconda.

The town of Anaconda was formed with homes, businesses

and hotels within three years and incorporated as a town. The two hotels that Marcus built for the town were the Florence, which was for the workers, and the Montana, a luxury hotel for visitors and townspeople to enjoy. Marcus decided that if Margaret thought this best for the family, that is what they would do. In Anaconda the Dalys built a three-story red brick Victorian house with seven bedrooms on the second floor and a ballroom on the third floor, with four servants' bedrooms on the north side of the ballroom.

The baby Harriot was one year old, Marcus II was three, Molly was nine, and Madge was twelve when the Daly family moved to Anaconda. This was Marcus Daly's town. This was Anaconda Copper Company's town. The atmosphere was friendly, and there was great respect for Marcus and his family.

Both Marcus and Margaret felt as if they could breathe more freely and that their heavy hearts were lightened. They felt that their new living arrangements would be much better for them; even New York didn't sound so terrible to Marcus. He was beginning to lean towards Margaret's way of thinking.

Chapter Four

The smelter in Anaconda was the largest in the world. Marcus tried to get the Northern Pacific to build a spur to Anaconda from Butte, but the line wasn't interested. Marcus had to have transportation from Anaconda, and he didn't want to use inefficient mule teams. Marcus built the railroad from Butte to Anaconda himself. Not only was ore shipped back and forth, but also lumber for the smelter was shipped in. The smelter used a large amount of logs in the furnaces to produce refined copper from the ore. The Anaconda mine in Butte used another large amount of timber, up to 40,000 board feet per day, to shore up the mines. As a result, the timber industry became a force in itself, and there was trouble about the regional forests being harvested too widely. Law suits by the government were being brought against the lumber companies. Marcus was part of the Montana Improvement Company that the government was suing. This was another worry for Marcus. Lumber was an essential product for the industry. Meanwhile, Marcus found it hard to believe that Clark cared more for political office than he did the betterment of the industry and still was battling on about politics. With the government accosting the practices of the lumber company, Marcus knew he would need to get new sources of lumber. Marcus thought of the Bitterroot Valley and all the timber on the west side of the valley on the Bitterroot Mountains. He knew there had been success with the apple orchards. He was interested in the orchards and the ranches which had sufficient natural grass for grazing, and where there was water, crops were good. The decision to buy land in the Bitterroot Valley was finalized in 1886. Marcus had agents buy seven ranches in their names, then he would buy the ranches from them later.

The Bitterroot Valley became a large enterprise with logging and a lumber mill. The Bitterroot River was a natural source for floating the logs as they were harvested, and a dam was built to keep the lumber in a basin until milled. One enterprise led to another, and soon Marcus was swamped with plans for a lumber mill. This industry needed a town. The town of Hamilton, Montana, was born to support the needs of the employees of the lumber industry of Anaconda Copper Company.

Each of the enterprises that Marcus Daly was involved in would be a 24 hour job for any one person. Nevertheless, Marcus was able to oversee all and manage astutely these business ventures. The Anaconda mine in Butte was very well organized and those in management positions were highly capable of carrying out the everyday operations. Major decisions or problems were reported to Marcus and solved by him. The smelter with the largest smoke stack in the world as well as the largest smelter, was in operation in record time.

Since Mr. Grant declined Marcus' offer to buy his town in the Bitterroot Valley, Marcus proceeded to build a town similar to Anaconda. This new town would support the lumber, ranching, farming and orchard enterprises in the valley. Marcus hired Mr. Jim Hamilton to be the engineer for the Daly Bitterroot Valley enterprises as well as for the Anaconda Copper Company. One of the many assignments he gave to Mr. Hamilton was to design a town with streets wide enough that a horse and buggy or mule wagon could make a complete turn around on the street. Mr. Hamilton was to coordinate land purchases and conduct surveys of them. He also was to oversee the building of the lumber mill, as well as the logging on the Bitterroot Mountains. As chief engineer, Mr. Hamilton needed a staff to accomplish his assignments, and Marcus gave him a free hand in hiring the men to work for him. When the time came to name the town, Marcus named the company town Hamilton, honoring Jim Hamilton and his accomplishments. As soon as Marcus had the master plan of the lumber industry under the care of Hamilton, he was free to devote his time to the Butte mine (now the Anaconda mine) and to the smelter in Anaconda, and to his family who were growing up before he could know them.

Anaconda and the town of Hamilton were laid out in the same design. Both towns had wide streets, spacious lots for building hotels for guests, a company store, a bank, churches, a city hall and houses for workers and management. They were company towns, Anaconda Copper Company towns. The workers were taken care of from cradle to grave and although they paid rent and bought their groceries, the buying power of Anaconda greatly reduced the costs of company store groceries and houses rentals. All of the amenities of the company towns were bonuses for their employees. The majority of the workers were able to live very well and to save money for themselves. Of course, there were always those who overextended their income and went into debt to the company store. Marcus was aware of this and would take the time to inform the offender to cut back on spending and get out of debt. If the debt was because of illness or some other misfortune, Marcus would give the debtor money to help the family back to solvency.

Marcus carried a little black book with his worker's names, as well as the names of wives and children, and he always asked about each of them. Loyalty to Marcus Daly was well known to all, and he had a solid base in friendship with the men in his employ.

Margaret had the home in Anaconda in order and the children settled in school in a short time. The younger two had their nanny to amuse and care for them. After assigning the routine of cleaning and arranging to the two girls who were maids and to the butler, Margaret had lunch and then retired to her bedroom to gather her thoughts. She needed to organize the duties for the other servants, plan menus for the week and decide what clothes to have made for the girls and Marcus II. A seamstress was needed. Margaret wondered if Mary Dolin, a seamstress from Butte, would consider moving her business to Anaconda. Margaret sent a note to Mary Dolin asking her if she would be interested in moving to Anaconda with the promise of a dressmaker's shop on Main Street.

"Dear Mary,

It has come to my attention that there is not a qualified and talented seamstress such as yourself in the town of Anaconda. I'm hoping you will consider moving here and establishing a seamstress shop. I know I could guarantee many clients and it could be a happy situation for you and the ladies of Anaconda.

Mr. Daly and I will finance your shop and you will stay at our home until your shop provides enough income for you to have your own quarters.

We do miss you and your fine sewing and hope this offer will persuade you to move to Anaconda.

Please answer as soon as possible as we are anxious to know!

Sincerely yours,
Mrs. Marcus Daly"

Margaret received a letter from Mary within the week.

"Dear Mrs. Daly,

I was pleased to receive your letter and your kind offer, which I accept with many thanks. I have always dreamed of my own shop, but I never thought it would be a possibility.

I will be in Anaconda at your front door a week from this Monday.

I have no family or ties to bid goodbye, and the shop where I work will have a replacement for me immediately.

Again, I thank you for helping me.

Sincerely yours,
Mary Dolin"

Margaret was delighted with the news and waited anxiously for Marcus to come home from the office so that she could tell him and they could proceed with Mary's shop.

Margaret heard the door open and called, "Marcus, dear, I'm in the living room."

"Margaret, me darling, you look all flushed. What has happened?"

"Mary Dolin is coming next week and wants to start the dressmaking shop you and I discussed."

"That's quick! I'll get some men to paint the building we talked about, then you and Mary can have the cabinet men build what you want."

"Oh wonderful, Marcus. Really, Mary is the best seamstress I've ever known. She is truly an artist. Thank you, dear."

"Glad to do it. I'm sure Mary will make a success of her shop. She is a fine woman."

Mary arrived as scheduled the following Monday. She was

short and small boned, but had a well padded figure. She had a round face, a short, pert nose, dancing blue eyes and a smile that angled to the side, giving her a quizzical look. She wore her steel-gray hair pulled back in a bun, with short bangs on her forehead. She was a most pleasant and merry person.

As she rapped at the door and was greeted by the butler, Mary never looked back as she walked into the parlor where Mrs. Daly was waiting for her.

"Mrs. Daly, how grand to see you again."

"We're so happy you are here, Mary. This will be a great venture for you and a wonderful addition to Anaconda. We have much to do getting your shop in order. But first, Mary, we must get you settled in your room."

"Oh glory, I love the room," said Mary.

"I'm so glad, Mary. You will have your meals with the tutor and nanny and children. Their hours are regular times and should fit your work schedule," said Margaret.

"Thank you, Mrs. Daly. This is more than kind. I don't know how I will ever repay you and Mr. Daly."

"You needn't worry, Mary. I have a selfish motive in asking you here. I needed a seamstress myself."

Mary Dolin's seamstress shop quickly became a success with some guidance and much patronage from Margaret Daly. Mary Dolin also imported and sold fabrics for the dresses she made for her clients. Mary was happy, and the ladies of Anaconda were happy.

Weeks later, with the day organized, Margaret let her mind wander and she thought of her time and the family's in Butte. So much had happened after the mine became a prosperous enterprise.

The greatest enjoyment for Margaret was that her mother, father, and three sisters had moved to Butte. The family gatherings were so joyous, and Butte society had accepted them all. They, in turn, had became leading citizens. The first years in Butte were wonderful with the W. A. Clark's accepting and introducing the Daly family to friends.

Margaret remembered the night her sister Miriam met W. A. Clark's brother, J. Ross. The two met and they never looked at anyone else that evening. Their heads were bent together in deep conversation, occasionally stopping when others interrupted. It wasn't very long before all of Butte knew that Miriam and J. Ross were a love match. A short engagement was followed by a formal wedding at the Episcopal Church in Butte.

It was one of the happiest moments for Margaret. She loved all her sisters, but Miriam and she were closest.

Miriam's wedding gown was white silk overlaid with embroidered lace dotted with pearls. Her bustle had lace ruffles cascading to the floor, and the veil was Brussels gauze lace held in place with a tiara of pearls. Of course the bride was beautiful with love. It was a perfect wedding and celebration.

Thoughts of the feud intruded on Margaret's reverie, and a frown passed across her forehead. It had become a nightmare. Marcus and W. A. Clark were engaged in this terrible battle, and Margaret and her sister no longer could meet openly, but were forced to meet secretly away from their husbands. J. Ross wasn't an enemy of Marcus, but W. A. was his brother and J. Ross couldn't ignore W. A. Before the feud it was truly a happy time. The children loved having aunts, uncles and grandparents at their command, and it was fun for everyone. Margaret recalled those wonderful times and decided that the good memories were what she would store in her mind and heart.

Coming back from her reverie, Margaret was startled to see it was 4 p.m. It was time for the children to be home from school and the little ones, Marcus II and Harriot, to have their snack. She decided she would go down to the dining room and have tea with her children. When Margaret entered the room to have tea with them there was a clamor for each one called, "Mama, sit by me." Margaret seated herself at the head of the table and asked each one about their day and she, in turn, told them of hers.

That day Marcus was to come home from his Bitterroot trip. Margaret wondered how Marcus was able to stand the pressure of his work and his many interests. She worried that he was burning himself out so early in life. He was just forty-five and had many burdens along with his many successes.

Margaret's parents and two sisters had followed the Daly's move to Anaconda and had a house a block from the Daly's. Margaret was sure this was a good move for her family. Butte had become impossible for any Daly connection as Clark was still using his newspaper to attack Marcus. Miriam and J. Ross could cope with W. A. Clark. Ross worked for W. A., and they, as a couple, were obligated not to remind him of their Daly connections.

In 1885 and 1886, Butte and Anaconda were becoming recognized by business men in the East. Boston investors found that copper was profitable; and Montana and -- particularity the Anaconda Company -- was making millions for the owners. Shrewd California businessmen decided Montana was a place to invest their money as well. The Montana Hotel in Anaconda was full of investors looking to find action in mining and evaluating the known properties.

George Lewis, from California, had been given a letter of introduction to Marcus Daly from the Wells Fargo Company, which was headed by Lloyd Tevis. Marcus was glad to meet George and asked about Lloyd Tevis, but George really did not know Lloyd Tevis personally. He had done some work for Wells Fargo; therefore, he'd been given the letter of introduction. Since it had been some time since Marcus had been home, he asked George Lewis to come home to meet his family and have dinner and a social evening. "Tomorrow," Marcus said, "will be time enough for business." Margaret had planned to have dinner with her sisters and mother and father, so one more didn't cause any trouble. Margaret was delighted that Marcus was coming home for dinner and not working late.

Hattie looked especially beautiful that evening. It was spring and she wore a cornflower blue dress with matching lace at the cuffs and neck and down the front in a "V". Hattie's hair was a rich auburn that set off the fine hazel eyes that glowed in her chiseled ivory face. Hattie was the merriest of the sisters, known for her sense of humor and good disposition. George Lewis had been very happy to accept Marcus Daly's dinner invitation because he wanted to learn more of the ways of this dynamic man. The bonus was the family, particularly Hattie Evans, and George appreciated both the meal and the excellent company. George had conversation with both Hattie and Mattie, but Hattie was the one he sought out after dinner. George would be in Anaconda at the Montana Hotel for a month, and Hattie noted this with pleasure.

"Miss Evans, what do you do with yourself here in Anaconda?" asked George.

"Nothing very exciting, but I keep busy. I play the piano, embroider and help my sister Mattie, who can't exert herself. She has a weak heart, you know. We have summer opera and theater in Anaconda and Butte, and we seem to dine with friends a great deal. Now, Mr. Lewis, tell me about yourself."

"Well, I'm connected with the Coastal Railroad in California. I'm an investor for a number of individuals and companies, therefore I travel frequently. The copper industry is a good investment, so I'm here making inquiries," said George.

Margaret was aware of the mutual attraction between their guest and her sister, and she had to smile to herself. With all the disruption, turbulence and tediousness of the last ten months, there was still wonderment in the finding of love. Margaret looked at Marcus and remembered and was able to share in spirit with her sister's new-found love. It was wonderful to have this joy and happiness invade the household.

The move to Anaconda was the right thing to do. Margaret and Marcus both remarked that the decision had been a godsend. Marcus hadn't had time to tell Margaret about the Bitterroot trip, but he had said that he planned to establish a lumber industry there and had agents acquiring land. "The valley is beautiful, Margaret. I would like to you to see it; and so you shall, as soon as we can get a week away," Marcus promised.

There was much warmth and happiness in Marcus' description of the valley. Margaret knew Marcus had found an interest that would make him happy and, as a result, she was happy, too. It had been a long day and after their visit and vows of love they both fell asleep. It was the custom to have separate rooms, and the Dalys did, but this night Marcus never left for his suite.

When the month had passed and George Lewis was to leave, he vowed his love to Hattie and asked Mr. Evans if he might marry his daughter. Everybody thought they were wonderfully well suited, and Mr. Evans gave them his blessings. When Hattie awoke the next morning she suddenly remembered the excitement of the engagement. She dressed quickly and readied herself for the day. She arrived in the dining room for breakfast, but it was such a beautiful day and her mind was in such a whirl, she decided to have her breakfast outside in the gazebo where she could be alone with her thoughts. The sweet peas were climbing the trellis and their fragrance sent Hattie into a world of her own, a new world with thoughts of her love for George.

"I'm engaged. I'm happy. I'm awed at the prospect of being a wife, but then there is George and I know that because I love him, his way will be my way. Isn't life a wonder with all its twists and turns? And I know this last turn is meant to be. I will be married to George!"

"Hattie, dear Hattie, where is your mind? I'm talking to you," said Mrs. Evans.

"Oh! Mother, I was thinking about getting married and about George," Hattie replied.

"That is all well and good dear, but you do need to get on with our life. You will need to go to the dressmaker, Mary, immediately and start on your trousseau. Thank goodness we at least have some beautiful linens hand embroidered and will have time to put your initials on the linens and towels. You can't imagine now much you and I will have to do. Of course, Margaret will help us. Have you told Margaret yet?" asked Mrs. Evans.

"Oh, no, I'm going to run right over and tell Margaret. I promise I'll be right back, then go to Mary's and look at patterns and materials. Won't Margaret and Marcus be surprised?"

With that, Hattie rushed into and then out of the house, leaving her mother shaking her head in wonder.

The wedding was to be in September at the Episcopal Church in Anaconda, and the reception for family and friends was to be in the ballroom of Marcus Daly's hotel, The Montana. September is perhaps the kindest month in Montana. The days are warm, the light is sharp and golden, and the nights are cool and clear. Color begins to dot the mountains as the aspens and larch start the process of shutting down business for the deadly cold days ahead, and the storied big sky is precisely the blue of the church's sanctuary where George and Hattie exchanged their wedding vows.

"What good luck", thought Margaret as she watched her graceful sister take George's hand. "Miriam's wedding dress fits Hattie perfectly and, though I'd never say it out loud, she looks even more beautiful in it than Miriam did!" As the groom kissed his bride and the old organ swelled, she smiled up at Marcus and silently renewed her own wedding vows in her heart.

The reception at the Montana Hotel, yet another Daly enterprise, was a lavish confection of gold and white. The colors were repeated in the snowy table linen festooned with gold garlands and ribbons, in the massive three-tiered wedding cake, and even in Margaret's gown, a chic beige silk with gold and white panels, white lace rutching, and gold beading that danced and glittered as it caught the light. An orchestra played Strauss waltzes earnestly as the company selected food from the extensive buffet that had been meticulously prepared and presented, a Margaret Daly trademark.

The reception even out-shone the wedding, which was spectacular with white and gold flowers and ribbons and gold candles. Hattie wore Miriam's wedding dress, for she had helped Miriam select it, and it was such a short time between weddings that the dress was still the latest in fashion and was as perfect for Hattie as it was for Miriam. Madge and Molly were big-eyed with excitement. This was their first adult social function. As flower girls in their gold silk dresses with wide taffeta sashes, they were expected to participate in all the festivities. However, Margaret drew the line at allowing them to have wine or to dance. They giggled on the sidelines as some of the heftier Daly friends attempted a quadrille. Smiling as she caught Molly's eye and shook her head "No" to all the girls' giggles, Margaret raised her hand to touch the pearls at her throat. It was a habit. Each time she wore the beautiful beads she was beset with a fear of losing them. Those perfectly matched pearls, as large as wild grapes, were a gift from Marcus, purchased from the first big payout of the Anaconda. She prized them even beyond their value because she knew what they had cost Marcus in terms of sleepless nights and 20-hour work days.

Margaret made her way through the crowd of guests to sit beside Mattie. She took Mattie's thin hands and smiled. "My dear, now that both Hattie and Miriam are gone away to be married ladies, you and I must stick together. Promise me that I'll see more of you. I know I shall be lonesome if you don't. Any time you want a guest for a good gossip, send me a note by that maid of yours, and I certainly expect to send the carriage for you whenever you're up to an excursion."

Mattie laughed, "Bless you, Margaret. I really needed to hear that from you! Frankly, I was feeling the tiniest bit sorry for myself." She kissed her sister warmly. "Of course, we'll see lots of each other."

Though Mattie's rheumatic heart caused her to be frail and limited as to what she was able to do, she was the sort of invalid that people liked and admired for her good nature and the cheerful, uncomplaining way she managed her disability. As Marcus was fond of saying, "Mattie may be a little slow out of the starting gate, but she's got the heart of a winner." On this day she looked like a winner as well. Her dress was very becoming to her. The dress was toast-color with copper and beige lace and braid cascading over the bustle in the back. Her hair was lighter than Hattie's but still had the reddish highlights. The ashes of roses silk bustled gown she wore was made to show her tiny waist and lovely skin to best advantage; and even though her health made dancing impossible, beneath the lacy hem of her gown one well shod foot tapped along with the music.

During the reception, Margaret looked at Mattie. Mattie looked frail and wan at the wedding and reception but always managed to smile brightly for her sister. Mattie truly was happy for Hattie and George. They would be missed, for it had been such a fun time for her, too, since George came into their life. Hattie and Mattie had become very close after Miriam was married. Not only were they friends and sisters, but also Mattie depended on Hattie for her well being.

Even climbing stairs was impossible for Mattie, so Evans house in Anaconda had been built with two downstairs to accommodate her. Mattie loved having Hattie pop in to her rooms for sisterly conversation. She would miss her visits.

Although lower necklines were in vogue, the Evans sisters kept the high necks on the bodice of their dresses. The shoes were of soft kid buttoned to the calf. They were comfortable, and the exact color of the dresses carried the theme of the autumn colors throughout. The Evans sisters were spectacular in looks and dress.

The newlyweds stayed at the Montana Hotel for their wedding night. The next day the bride and groom had brunch at the Daly residence and Marcus was the host and set the scene for a high-spirited farewell for the newlyweds. Again the board was laden with ham, sausage, baked eggs on toast, canned fruits from the Daly larder, hot chocolate, tea, coffee, and potatoes oven-baked with butter and cheese. Another feast and all were able to eat heartily.

Marcus had a serious moment and toasted the couple with a prayer. "I said a prayer for you today and know God must have heard. I asked that He be near you at the start of each new day. I asked for happiness for you in all things, great and small. But it was for His loving care I prayed the most of all."

This startled the group, for Marcus never was this serious with friends and family. All were touched and, instead of a toast, all set their glasses down and in unison said, "Amen." Then Marcus started the conversation and festivity again, saying, "Eat, drink and be merry!"

The train didn't leave until evening, but the Lewises wanted to be early, so they left in the afternoon in the Daly's carriage driven with a matched team of bays. All were waving goodbye to the newlyweds and throwing rice for good luck. They watched until the carriage was out of sight and Hattie's wave could no longer be seen. The next day the newlyweds boarded the train at Butte and went on to San Francisco and their new home.

By the time the newlyweds stopped at the hotel and packed their luggage, there was not much wait at the station. It was not long before the train arrived and they were boarded and on their way to Butte to connect with the Northern Pacific.

Mr. and Mrs. Evans and Mattie were the last to turn away. Mrs. Evans had been encouraged to wear a light copper-colored gown and, although it was not the color she thought she would like, it was a becoming shade for her auburn hair, which was streaked with silver. Mrs. Evans' features were not as soft as her daughters. She was an imposing woman and wore a piney for her eyesight. Her habit of putting in the piney and then looking straight into a person's eyes unnerved many a young beau. Mr. Evans touched Mrs. Evans's arm and gave Mattie his arm to walk back into the solarium and bid their goodbyes to the gathering. Now that train travel was possible, Margaret knew that she would be able to see her sister at frequent intervals. Although some tears were shed when they parted, they were not tears of unhappiness. It was a joyous occasion, with good friends and family. It would take a few days to settle into the familiar routine.

Margaret was concerned about her father. He seemed to be short of breath. When asked, Mr. Evans would reply, "The altitude is finally getting to me after these many years."

Madge and Molly were ten and twelve years old and started school after Labor Day in September. They were eager to start a new life with new friends and some old ones in a new town. On their return from the first day at school, Madge and Molly burst into the house calling, "Mama, Mama it was wonderful! We're going to have a beautiful geography book, art classes and poetry. Isn't that wonderful, Mama?" And, Margaret was content to know their well-being had been accomplished with the move to Anaconda. The routine of school for the two older girls and activities in the household by Marcus II and Harriot kept all the servants and Margaret busy.

There was no peace for Marcus. The Michigan mine owners were the only copper producers in the United States for many years, and they were not eager to have competition. Since they had ample resources, they decided to eliminate the competition with a price war. Anaconda Mining Company met the challenger where others failed, bringing the Michigan mine owners to their knees.

An interesting item that appeared in the newspaper about copper mining was brought to the attention of a Frenchman in Paris. His name was Hyacinthe Secretan, and he had cornered the silver market when the price was depressed. He thought the same could be done with copper. He knew the price was depressed and the demand for copper was increasing. Why not corner the copper market?

Secretan formed a syndicate with sixteen European investors and approached copper producers with a price fix of thirteen cents for the copper they produced instead of the ten cents they were now getting. He was immediately successful in signing up many of the copper mine owners.

Marcus was very troubled by this syndicate for he did not like to curtail production. He discussed the issue with Ben Haggin, who had become the decision-maker for George Hearst and Lloyd Tevis. Haggin was in favor of the syndicate and Marcus could not sway him. The guaranteed thirteen cents a pound for copper would add significantly to the company's profit. Marcus watched as the syndicate bought copper and set the price at thirteen cents. The companies using copper balked at paying thirteen cents and started using scrap copper. This was the ruination of the syndicate in Europe, and Marcus thought it was good riddance. This was a three-year episode, lasting from 1885 to 1888. Nevertheless, Montana had been discovered by the international business community; and the copper industry, particularly the Anaconda Company, caught the attention of the Standard Oil Company Trust. Marcus was aware of this interest, for he had made many trips to New York.

Margaret had not forgotten her desire that their children be schooled in New York. Again she tried to discuss the move to New York for the school year with Marcus. "Marcus, please think about a New York apartment so that the children's education can be enhanced. You are in and out of New York all the time. Why can't you live there and travel to Butte instead of vice versa? It means a great deal to me, Marcus. I know this would be best."

"Margaret, I know how you feel, but I don't think I will like New York," said Marcus.

"How do you know? Let's try it."

Marcus finally relented. He was sick of the same argument about New York, and he also thought he might be wrong. Maybe it would be fine.

The Dalys looked at Margaret's fine selection for a New York residence in the fall of 1886; the address was 725 Fifth Avenue. The brownstone was fashionable and Margaret wanted to furnish the apartment for the family in French Louis XV style. The purchase was quickly completed and Margaret was happier then she had been in years.

Upon Marcus' return to Anaconda he met with his old friends at the Montana Hotel saloon. "Well me lads, I'm tired, I need a quieter life, my wife wants to live in New York for the betterment of our children's education, and I'll be moving from Anaconda. To me 'tis a sad day, laddies. But, my friends, I will be back to see to the mines and the smelter and to see you," said Marcus.

The friends didn't know how to react to this announcement. Finally, one of the men said, "Marcus, we wish you good luck in New York and understand your leaving; but we will miss you and look for you to return anytime you can."

"Thanks, boys, I'll remember. Let's drink to that. Irish whiskey for everyone, a toast to my new life," said Marcus.

On arrival in New York, the family entered the five-story brownstone building at 725 Fifth Avenue. The vestibule had a glass and gold trimmed cage elevator. Molly said, "Oh papa, look at the fairyland elevator. Isn't it beautiful?"

"Yes, it is, darlin'. Let's see if the beauty works," said Marcus. They all got in the elevator and Marcus started the process by pushing the lever to "go" and stopping it at the first floor by returning the lever to "stop".

"Almost like steering a ship," said Marcus. As they stepped off the elevator, their butler came to greet them and informed them their luggage was in their rooms. The first floor consisted of the living and dining rooms with a study and butler's pantry. The servants had quarters on the lower level where the kitchen and larder were. The next two floors were for the family, with guest bedrooms and sitting rooms. Molly and Madge were excited about their rooms and went to find them.

Margaret called to them, "Girls, you didn't notice our beautiful rooms here. The decorators have worked months to accomplish this."

"Everything is pretty, Mother. I like the French furniture. It's different," said Madge. Molly echoed Madge, saying, "Different but nice."

While the rest of the world was decorating their homes in heavy Victorian, the wealthy of New York were importing antiques, especially furniture in the Louis XV fashion. The Savonnerie and Oriental carpets were laid on the wood floors. The furniture was of smaller proportions and more delicate. The chairs featured rounded backs, and there was tufted upholstery on the furniture that was covered in silk damask and embroidered silks in rich green, burgundy and gold. Belgium tapestries and antique paintings had been selected and placed tastefully throughout the living and dining room. The overall effect was rich but light decor.

Margaret was delighted with the finished result, and the children thought it all pretty. Marcus was not comfortable with such luxurious furnishings, but if that was what Margaret wanted, he would not spoil her pleasure.

After three months in New York, Marcus asked Margaret to meet with him in his study to discuss some future plans. Margaret thought Marcus would discuss building a summer place -- maybe even in Newport, Rhode Island, where the society people of New York spent the summer. An excitement welled up in Margaret and she wished that dinner was over and the children were settled for the evening. She couldn't wait to talk to Marcus and tell him how happy this move had made her. Margaret also wanted to ask him for a new elevator for the building. The latest elevator was run by electricity and much safer. She would like their home to be one of the first to have the electric elevator. The glass cage would still be there in its magical beauty.

When Margaret entered the study, Marcus looked up from the desk; she could see that he looked sad and her spirits sank. Marcus asked Margaret to come sit across from him so he could explain his plans to her.

"Margaret, dear heart," said Marcus, "I know you are happy here, I know the apartment is to your liking, and I know the children will have many advantages; but New York holds nothing for me except our family. I cannot live here with my business interests in Montana. I must look after me laddies, for what will happen to them if I'm not there to fight for their rights? Standard Oil Company isn't people-oriented like my Anaconda Company, and Standard Oil is going to be the majority stock holder. As president, I will have some power and look to having a fight to retain my ways.

"Besides, I'm really not socially acceptable here. As an Irishman, I'm a minority and a liability in this social scene. You and the children can be accepted, but I must have my life in Montana and, of course, with you. The visits here in New York and the family coming with you in the summer to Montana will keep us together."

Margaret couldn't believe what he was saying to her. What would she do without his support? Her eyes filled with tears when she realized he felt so inadequate socially. Marcus calmed Margaret, dried her tears and let her know he wasn't deserting her. Train travel was adequate and he could be at her side in a few days if needed.

So, Marcus returned to Anaconda. His friends and workers were glad to see him, but asked why he changed his mind. Marcus's reply was, "I have nothing in common with the New York businessman, and I hate the French food they serve. Actually, I missed you laddies."

Marcus moved into his suite in the Hotel Montana. He was a plain man and his room was spartan, with its oak bedroom furniture, a sitting room with comfortable but simple furniture, and a desk with a matching chair. This style suited him.

In her correspondence to Marcus, Margaret mentioned a summer home and told him about Newport, Rhode Island. Marcus knew he would hate that. Actually, the only thing he liked about New York was the thoroughbred horse races at Belmont. Margaret let it be known that Anaconda wasn't suitable for a summer place. Marcus agreed it would not be a place for him to have freedom, as this was an everyday place to him.

Marcus was acquiring a great deal of land in the Bitterroot Valley, and though he was having a large home built in Hamilton proper, if Margaret wanted a large summer home with grounds, the recently purchased Chaffin place would be ideal.

Marcus couldn't wait to write and tell Margaret about the prospect.

"Dear Margaret,

I have just finished the purchase of seven ranches here in the Bitterroot. One of the ranches, Chaffin, has a fine mansion which you could enlarge and make into a home as pretty as you desire. The main stock farm will be adjacent to the mansion, and the rest of the 28,000 acres will be divided into six more farms. One will be a dairy farm that produces cheese and butter as well as milk; another farm will be for grains and alfalfa; and another will be for nursery stock (plants, trees, grasses, etc.). Of course, another will become a horse farm for working, riding and carriage horses, and breeding stock. The thoroughbred horse farm will be my special interest, and I can't wait to start on the quest for horses for the Daly stable!

All in all, Margaret, the Riverside Estate with its seven farms and the Daly Mansion as its centerpiece jewel will fulfill both your and my dreams.

Love you, Margaret. Hope this finds all is well with you and the children. I'll see you soon.

Your faithful husband,

Marcus"

Marcus was especially excited about the thoroughbred farm. He would have the finest race horses and hoped to challenge the East coast horses at Belmont. He had decided that the higher altitude in the West would enlarge the horses lungs, and they would have more oxygen to run. Marcus believed the timothy hay and natural grasses were more nutritious in the Bitterroot with the drier atmosphere.

Marcus knew Margaret would accept his plan. It would be a kingdom, and she could invite all the family and friends to spend the summer season right in their home. He would see to it that the Mansion would be large enough to accommodate everyone.

Marcus decided to just go to New York and tell Margaret the plans. This enterprise was reviving Marcus and his spirits lifted, for he had been battered by his ongoing feud with Clark, the acquiring of Anaconda stock by Standard Oil Company and the work load of his businesses. Aboard the train he drew plans to show Margaret. He also would order another private car for the family to use, and he would build a track and a railroad to his new town, with a loading area at Riverside.

Marcus arrived in New York and was met by the Daly town carriage. After the loading of cases and Marcus himself, the horses and carriage went quickly to 725 Fifth Avenue where Margaret and the children were waiting. It was at 5:00 in the evening that Margaret heard the elevator start. She knew that would be Marcus. The door to her sitting room opened, and Marcus rapped on the side as he walked through. "Margaret, me girl, where are you? I have so much to tell you."

Margaret answered, "In here, Marcus, I'm coming to the sitting room. It is wonderful to know you are here. Was the train late? Did you ride the Daly railroad car? Are you tired? Did you leave everyone well?"

"Margaret, wait a minute. I can't answer all from this long distance. Come here and greet me with your lovely smile and a sweet embrace. There, my dear, now I will tell you everything."

Marcus and Margaret sat together on the divan, and he held her hand, playing with her wedding band and twisting it on her finger.

"Margaret, not only have I been successful in acquiring the acreage to have a summer home as large and grand as you want, but also I've purchased 28,000 acres surrounding the house and intend to have a self-sufficient operation. I want the farms to be productive as well as financially responsible. They should pay for themselves, or perhaps even turn a profit."

Margaret said, "Stop. Marcus, where did you buy the land?"

"About one mile outside of Hamilton proper. You can ride into town with horse and buggy and not take all day. The Chaffin house is there -- and a fine home it is -- but I know you will want to enlarge and add on to the house and put your touch on it, darling."

"Oh Marcus, it sounds wonderful, but what of the house you built in Hamilton?"

"Not to worry, my dear, that building will be an office for Mr. Hamilton and the Bitterroot stock farm business."

After Margaret and Marcus talked, Marcus started calling for the children to come and join him and their mother in his study before dinner. They all rushed to him with hugs and kisses, and happiness reigned. Standing aside, Margaret could stand it no longer and walked over and put her arm through Marcus' and patted his hand while he finished his greeting to the family. Dinner was a spirited affair that evening, even Harriot at three years was included in this family evening.

After dinner when the household had retired, Marcus and Margaret had their time alone, and they talked of their plans for a summer home. Margaret was overwhelmed with the plans Marcus had made for life in the Bitterroot: a mansion for her to design, surrounded by beauty with mountain vistas in every direction, and 28,000 acres of productive land. "A private resort," Margaret thought to herself. The family could gather and spend their summers at Riverside. It was a dream, but it would be a reality, for Margaret knew Marcus, and what he accomplished was always phenomenal. She doubted not, and told Marcus it was a wonderful

gift from him to his family and friends. So Riverside was born as an idea, and by the year 1888, the old Chaffin house was incorporated into a Victorian Mansion with bedrooms for everyone.
By the middle of September 1887, Marcus had acquired the land from his agents for the Riverside estate. He hired a foreman for each farm, and the foreman hired the local men for the labor. Overall supervision of the seven farms was the responsibility of the manager. One exception to this arrangement was the thoroughbred horse farm. The thoroughbred horse farm would be an extensive operation. Beside raising thoroughbred race horses, there would be an established string of thoroughbred trotters.

While the Riverside farms were being established, Margaret sent plans for the Victorian house incorporating the already established Chaffin house. Margaret could not abide waste, and the Chaffin house was a perfectly respectable Victorian house that had been built just seven years before. The architect in New York took the Chaffin plans that Marcus had sent to Margaret and added on the requirements set by the number of visitors and the luxuries that Margaret wanted to have in the new house.

Margaret realized that this summer residence could be as exciting as the summer homes that the New Yorkers had in Newport or in Maine. Knowing Marcus would be happy at Riverside was another great comfort for Margaret. Since they were separated part of the year, Riverside would be appreciated by both Marcus and Margaret as well as all the children. Marcus received the house plans in late October and called his engineer, Mr. Hamilton, to add one more duty to his responsibilities. Mr. Hamilton was to subcontract the labor and oversee the construction of the Daly Mansion at Riverside.

The Chaffin home faced west with a view of the Bitterroot Mountains that was breathtaking. To incorporate and enlarge upon the house, the front would have to face east with a view of the Sapphire Mountains. This was not a problem because either panoramic view was an inspiring vista. The living and dining room of the Chaffin house became the kitchen and butler's pantry. As a result, the kitchen had large windows and important door knobs and paneling which was not done in the kitchens of that period. The childrens' dining room and the servants' dining room, with adjoining storage and cooler room, made up the first floor of the original house, while upstairs there were three bedrooms. The third floor had a sewing room and servants' quarters.

To this Chaffin house was added the Victorian House, making in total twenty-four bedrooms, fifteen baths and closets, a living room, dining room, music room, office and an entrance hall with a grand staircase to the second floor. The second floor had twelve bedrooms and eight bathrooms, with a sitting room in the center. The first floor had two bedrooms and two baths as well as the public living areas. The third floor had six servants' rooms and two baths on one side of the ballroom. The other side of the ballroom had four bedrooms and three baths. The exterior was wood siding painted a soft green, with copper and yellow trim around doors and on the posts of the wraparound porch. Landscaping was immediate, as the nursery farm had received trees and shrubs and bulbs in the summer of 1886 that were ready to be planted at the Daly mansion in the spring of 1888.

The first year in New York had been lonely for Margaret, but the schools were superior for the children and the people they associated with were from very fine families. Margaret was content with her decisions. Marcus came to New York frequently for Anaconda Copper Company business, allowing him to be a part-time husband and father during the winter months.

While in New York, Marcus made several trips to thoroughbred farms in Maryland. After observing these, he knew what he needed to do to plan the thoroughbred farm at Riverside.

Margaret's mother and father, Mr. and Mrs. Evans, lived in the house Margaret and Marcus had built in Anaconda. Marcus kept his quarters at the Montana Hotel. Anaconda would be the first stop for the Daly family on their arrival from New York, and Grandmother and Grandfather Evans' hearts overflowed with love in anticipation of the family's arrival. Marcus was due back from New York after being with the children and Margaret. Marcus arrived in Anaconda from New York and went immediately to the Evans' home and rapped at the door of his own home.

Mary, the faithful parlor maid, answered the door and welcomed Marcus.

"Hello, Mary, are the Evanses still up and able to see me?"

"Of course, Mr. Daly, they've been waiting all day knowing you were bringing them news of their loved ones."

Mrs. Evans, a tall woman wearing severe black clothes, but not a severe manner, came rustling down the hall to greet Marcus. "Marcus, my dear, come in and sit with us, We are longing to know all the news of Margaret and the children. Are they well? Have Marcus II and Harriot grown? Is Margaret well and happy? And Molly and Madge -- are they happy being in the city as young adults?"

As soon as they entered the parlor, Mr. Evans rose and come toward Marcus and shook his hand with vigor. "Marcus,

Marcus, we've been waiting for you and appreciate your coming right to us with your news. How did Margaret like the Riverside Estate plans? Won't it be wonderful for us all to be together for at least four months each year? It will make our living apart from the family bearable. Summer at Riverside! Marcus, I believe you have done something that will bring happiness to us all."

Marcus replied, "Margaret is happy with the Riverside plans and, as of now, is busy with her ideas. She will certainly keep the architect busy devising plans. I myself have many plans for the different farms and will be busy the next several years planning and purchasing. It is a great comfort for me to know that you are here in Anaconda for my family."

After a thorough review of the news, Marcus said good night to the Evanses. The Montana was a fine hotel, and on this evening it was filled with guests. Marcus was lucky that they hadn't rented his room, as they did on one occasion. "Perhaps they have learned to keep my room free when I'm due in," Marcus said under his breath.

At least the desk clerk knew Marcus and greeted him. "Mr. Daly, your room is ready, and the dining room will send up fruit, bread and cheese for your bed-time snack."

Marcus thanked him, deciding he had thought, talked and planned as much as he wanted for that day. He ate an apple, sliced a piece of cheese and drank a bit of Irish whiskey. To go to bed was what Marcus wanted. In less than a minute he was asleep. Marcus' ability to sleep anywhere, anytime had always been a great blessing. Business, problems and family were all curtailed. Sleep was immediate.

The next morning Marcus started on his plans for the

thoroughbred farm. Marcus had contacted Sam Lucas, a trainer at Belmont Stables who seemed interested in Marcus's theory that the higher altitude would cause the horses' lungs to develop a larger capacity. The end result would be horses with more endurance to race the longer courses. Sam Lucas had not said he would be the Daly stable trainer, but he hadn't said he wouldn't, either. With that thought, Marcus went to the telegraph office to send Lucas an offer that he couldn't refuse.

Sam Lucas was a successful trainer and a good judge of horses. A trainer's advice was essential to the building of a winning stable. Marcus was relieved when a telegram was delivered to him at the Montana Hotel. Lucas had accepted the offer. He would be the trainer for the Daly stable. If Sam Lucas had realized what an enterprise this would be, perhaps he would have said "No." Although the monetary reward was higher than any trainer had received, and the selection and training of the horses was a huge responsibility, Lucas was to have much more on his plate. The horses were to be racing at Anaconda and Butte and eventually at Belmont. The logistics of getting fifteen thoroughbreds on private railroad cars, with Bitterroot grown timothy hay, two veterinarians, several assistant trainers, numerous jockeys and fifteen stable boys to help on the long journey was a challenging task.

Marcus had a fine trainer in Sam Lucas. Sam was able to get the thoroughbred farm and facilities in production quickly. Horses were essential and Marcus bought every winning horse at the races that was for sale. In the following years he purchased the finest stock in Ireland, England, Germany and France, as well as the United States.

Above the dairy farm on a lane south of the mansion was a

hill. This was the area that Marcus and Sam Lucas thought would be best for the buildings needed for the horse farm. Looking down on the valley and across at the Bitterroots was the vista that Sam wanted for his house. Across the road the veterinarian would have a house, and further south would be the veterinarian hospital equipped to specifications of the vet. Close to the houses, but far enough away for privacy, the jockeys' house, workers' bunkhouse and central cookhouse were built. The cookhouse became the social club for the jockeys and stable boys, where checkers and cards were played until "Cookie" closed the door on the dining room and on their games. It was a joyous life and excitement was in the air as the men anticipated the great beginnings of Marcus Daly's phenomenal rise in the thoroughbred racing industry.

Each of the other six farms had foreman's houses and bunkhouses and a central cookhouse.

As each farm became successful and had surplus, the produce was shipped and sold to Butte, Anaconda and Hamilton. The stock farm with the manager's house was adjacent to the mansion. The cattle grazed over the thousands of acres and went on to the summer range on the adjacent Sapphire Mountains.

While the Daly empire was being formed in the valley, the town of Hamilton was founded. The logging of the Bitterroot Mountain range was employing large numbers of men. A lumber mill was needed and built. The logs were brought down from the mountains by teams of horses straining and the driver braking to keep the wagon from running away as they came crashing down the mountainside. The logs were dumped unceremoniously in the river to float down to the dam at Hamilton and into a holding basin. It was dangerous work, and miraculously more horses and men were not killed or maimed. The Anaconda mine and the Anaconda Smelter were voracious users of wood. The west side was harvested up to 5000 feet. The lumber industry and support labor made Hamilton a thriving industrial town.

Sam Lucas came to the Bitterroot and Riverside in the spring of 1890. He was embarking on a career that would take him far beyond a trainer's job. At his first meeting with J. L. Hamilton, they discussed Marcus Daly. J. L. said, "Sam, you won't believe what you can accomplish. Marcus Daly will expect great things and will support you on projects you never thought possible to bring to a conclusion. But remember one thing: Marcus never asks you to work harder than he does himself. Marcus is a visionary, and we all follow his dream. Now he wants me to work up a system to irrigate the entire Bitterroot Valley. I've already dammed Skalkaho Creek, and Marcus's farms at Riverside are irrigated. This Lake Como project will take some time and plans, but it can irrigate the entire valley from Darby to Stevensville."

"That is amazing," said Sam. "I plan to keep everything under control as much as possible. But, today Mr. Daly said he would like to race in New York at Belmont this year. I'm wondering if he is serious."

"I can guarantee he is serious, and you'll accomplish it just as you will many more things," said J. L. "Let's go over to the Ravalli Hotel and have lunch. We may not see each other for a while if we start on these projects." They laughed and went out of the stock farm headquarters to lunch at the Ravalli Hotel.

The hotel was built to take care of all the people coming to look at orchard land. The apples grown in the Hamilton-Victor area were the finest McIntosh apples, sweet and juicy. With the growth of Hamilton, the Ravalli Hotel became the social center for parties. An opera house was built on Main Street. The north side of Hamilton was lined with houses for workers. The south side of Main Street was for the more opulent managers' homes.

Many Finnish people came to work in the lumber mill. The Finnish were a very clannish lot and preferred that their wives and children not associate with others. They built their own clubhouse and established a social life among their own.

The life in Hamilton and the life at Riverside were separated by two miles, but the details of those lives were leagues apart.

Chapter Five

In the spring of 1888, the Riverside mansion was finished. As soon as school was out in late May, the Daly family would occupy their summer home.

Margaret was talking with the children at afternoon tea.

Margaret said, "Pay attention, children, I have exciting news of our summer home in Montana. We will be leaving soon on the train. You need not worry about packing your belongings. I will send the summer clothes with the servants who are going early to make the house ready. We have invited friends to join us for the summer. You will see Dr. Murray and his family, as well as lots of other friends who will come and go during the summer. You each can have a special friend for a visit, as we have plenty of room. There will be horseback riding, hunting, fishing and hiking, as well as games on the grounds. It is going to be grand, and we will be an extended family including friends and relatives. Your father has provided us with a gracious summer home in the beautiful Bitterroot Valley. I know you will be pleased and happy. You must thank your father for always giving you the best of everything."

Six-year-old Marcus II was the first to speak. "I don't want to leave my friends here. I don't even know anybody in the Bitterroot."

Then Harriot chimed in with, "Me too, me too!"

Margaret promised Harriot, "There will be many horses to ride and you may choose one for your very own." This won Harriot's heart, as she loved horses passionately.

"Oh, wonderful, Mama. Horses are my special friends. Can I ride as much as I want?"

Marcus II was not so easily appeased until Margaret said,

"Marcus, you can hunt and fish all you want. The valley is overflowing with game and fish. You can become an authority on the game birds, animals and fish in the Bitterroots. Guides will take you out with them to observe, fish and learn to hunt. Our life at Riverside will be of our own liking and desires."

As the conversation continued, they all realized this was a privileged life that few people had experienced. Even their wealthy friends who summered in Newport or in Maine would not have the freedom, the variety and the luxury of a kingdom of their own. The older girls, Molly and Madge, agreed.

Madge said, "Mother it will be wonderful to see our old friends and be with Grandfather and Grandmother Evans, Aunt Mattie, Aunt Hattie and Uncle George. I really have missed them while in New York. Mother, you know I like to ride horses, but most of all I like to drive a carriage. Do you think I'll be able to do that?"

"Yes, Madge, I'm sure that can be arranged if your father thinks you are capable," said Margaret.

What a relief for Margaret to know that the young people were as pleased as she and Marcus were. It was a happy day at 725 Fifth Avenue, that April 20, 1888.

Margaret found it difficult to let her servants leave early, for they were the mainstay of her household. She decided not to fret over the New York apartment household replacements, for they would mainly be caretakers of the Fifth Avenue apartment while the family summered at Riverside.

The time came for the servants to leave. Mr. and Mrs. Cutler, Stena the Swedish cook (who was a kitchen tyrant but a heavenly pastry cook), and Eddie the black handy man, well versed in serving or butlering as needed would set up the household at Riverside two weeks before the family arrived.

Margaret rang for Cutler the day before departure, "Cutler, I'm depending on you to instruct the others and supervise their work. I know that you can be depended upon and that Mrs. Cutler is a great help to you. Of course, if there are some decisions you are not comfortable with, I will take care of them when I arrive, but I have faith in your judgment."

Cutler replied, "Thank you, Madam, I will do my best. Since Mrs. Cutler is a perfectionist, all should be ready for your inspection when you arrive."

As Cutler was leaving, Margaret asked him to send Stena in to receive instructions. "Good morning, Stena. Are you ready to make the journey and organize a new kitchen? You will have as many young women as you need to help you, but you are in absolute charge of the kitchen. I will give you the menus two days ahead, and, of course, as always, I will rely on you for suggestions. I hope you are looking forward to our new adventure."

"Oh yah, Madam, it will be so good to be in the country and away from the summer heat in New York. I'm anxious to be on the way."

"Stena, Mr. Cutler will supervise the household staff and will be interviewing and hiring servants. I want you to be there when he interviews for kitchen help. You know who you can work with, and you are a better judge of cooking skills."

"Oh yah, Madam, worry not. I will be there if it has to do with my kitchen. Thank you, Madam, and all will be ready for you. Tell the children I will have the Swedish sugar cookies ready when they arrive. They all love my Swedish cookies."

"We all like your pastries, Stena, and your fame has spread in New York City. Now, the Riverside household guests and relatives will be treated to your talent."

Eddie was called next. He would be in charge of seeing that the trunks were placed in the proper bedrooms, that everything they contained would be placed in the proper closets and drawers. "Eddie," said Margaret, "you will look to Mr. Cutler for your instructions just as you do now. I've spoken to Mr. Cutler and he will put you in charge of luggage and placing everyone's belongings in the proper rooms. I have given Mr. Cutler a floor plan and designated each person's room, and he will help you identify these. I think you will like the West, Eddie. It is another world."

Eddie replied. "Yes ma'am, Mr. Cutler has been telling me what Mr. Daly told him about the Bitterroot Mountains, big and craggy, and the Bitterroot River, clear and cool with many fish, beavers and muskrats. I've never seen anything like they describe and I thank you for letting me go west."

The four family retainers were ready to go and follow Mr. Daly's dream that was already a reality.

Cutler embodied the English Butler stereotype. He was tall and fair with a clipped accent and a kind smile for the Daly children. He knew Mr. Marcus Daly's story and was frankly amazed that Mr. Daly would have an Englishman in his home, given the history of conflict between the Irish and the Black and Tans. But typically Marcus judged Cutler as an individual quite apart from his origins.

Mrs. Cutler was an exuberant woman who tried to be sedate when her husband was around, as he was quick with a cold stare of disapproval when he thought anyone was being undignified but the rest of the help liked her jolliness and hearty laugh. Cutler said Mrs. Cutler was a perfectionist, but actually it was he who could not abide imperfection. Since Mrs. Cutler tried to please her husband, she became a perfectionist by association. She was a little plump, with reddish blonde hair, round blue eyes and arched brows that gave her a surprised expression. Probably Mrs. Cutler was the favorite of all, for Cutler was all business, Stena was all temperament and Eddie was helpful to all.

The time arrived and the four family servants boarded the pullman car. Other cars provided with the family possessions were joined to the train.

The furniture that had been selected in New York was in place at Riverside before the Dalys departed Fifth Avenue. Cutler had the plans for the furniture placement in the mansion according to the designers. Mr. Daly and the servants made the journey in the private Daly railroad cars. This was reassuring to the servants since the journey's end was an unknown element.

The New York Central Railroad system operated the trains from New York to Chicago through the heartland of the country. The passengers saw rich farms with good black soil producing bumper crops of corn, wheat, barley and oats. From Chicago through the northern states, the Burlington and Northern Pacific operated. The Daly private car and the pullman were sidetracked at Chicago and attached to the train of the Burlington Railroad and later the Northern Pacific. Here the train traveled through more farm lands and the hills of Wisconsin, lakes and farms of Minnesota, and the plains and flatlands of the Dakota territories into the Montana territory to Missoula, Montana.

The trip was quite an adventure for the domestic staff. As

they left the crowded city of New York and entered into the farmlands of the midwest, the cities were few and far between. Chicago was the largest city to be seen since leaving New York. Since the extent of the servants' travel had not been over 30 miles outside of New York, they were craning their necks so as not to miss any sights. The land was rich and the farms looked prosperous.

Mr. Cutler said, "Isn't it hard to believe that all this farmland we've been traveling through was once forested? Imagine clearing all the timber to make these farms."

This was not such an adventure for Marcus Daly, who frequently made this same trip to and from New York.

Outside of Butte, Montana, the town of Garrison was a train-switching station. The Daly cars were switched to the Northern Pacific line going into Missoula. Marcus called Cutler to his car and said, "Cutler, the staff will be wondering what is happening when the cars are sidetracked for a short period before we hook up to the train going to Spokane. Reassure them that this is a regular procedure and not an emergency."

"Yes, sir, Mr. Daly. I will speak to them about the situation. It will help me to keep them from worrying. They are looking for Indians and are concerned about the tales they're heard of the wild west."

"Well, Cutler, it wasn't too many years ago we were still having the Indian wars, but now it is train robbers and highway men that you need to worry about. I carry my pistol but have not had to use it. We are getting more civilized here in the West."

On arrival in Missoula, the Daly cars again were sidetracked. The Northern Pacific had built a spur into the Bitterroot Valley. The tracks were only laid as far as Riverside. This was fortunate, as a stage journey from Missoula to Riverside was a slow tortuous trail that followed the wandering Bitterroot River. The journey had previously been a two-and-a-half-day trip. Now, the train made it in a half day.

As the train neared Riverside Estate, Mr. Cutler caught sight of the Daly Mansion. "Stena," said Mr. Cutler, "look, that is our grand home. The Mansion looks as if it is a beacon all alone in this vast land. It is our safe haven."

Stena then saw the carriages and wagons waiting on the lane leading to the Mansion. "Mr. Cutler, there they are waiting for us. When you see this wild country you wonder what will be next but by glory, they are here to meet us after a 2,500 mile journey. That's amazing! Here we are at Riverside."

Mr. Daly took the first carriage and called to Cutler, "You go ahead and do your best; I'm going to town and will be there until tomorrow. I'm at the stock farm office in Hamilton."

Mr. and Mrs. Cutler and Stena went off in the second carriage and Eddie stayed to help unload the railroad cars and take care of the trappings.

As they drove up the lane lined with small but sturdy trees, Mrs. Cutler could contain herself no longer. "Isn't it the most beautiful and wondrous country you have ever seen? Those big mountains across the way take my breath away. They are so craggy and you can see in between the canyons that the mountains go on and on."

"My, we will never be the same again, Stena, for we have seen some God-made wonders," said Mr. Cutler. The closer they got to the mansion the more their excitement grew. Even Mr. Cutler said, "The mansion is glorious — a grand summer home."

The carriage drove to the back entrance with its wide steps and full porch facing the Bitterroot Mountains like a front entrance. The threesome climbed down from the carriage and entered the back hall. The servants' staircase to the second floor was directly ahead. To the side was a pantry with walk-in ice box and cooler. To the right they entered a large kitchen. Stena exclaimed, "That's the stove that is the best — ah wunderbar — we are as well set up as we can be, just like in New York."

Just then Mr. Cutler came into the room. "I never thought we would have so complete a serving area and kitchen. There will be no more work here than in New York. We even have electric lights! They must be the first in this part of the world. Mr. Daly told me he had his own generators and that the Daly summer home would be modern, but this is amazing. Let us take a walk through and see the rest and accommodations. I have the floor plan so we will know who and what belongs in each room."

As they walked through the house they found each room furnished and ready for their care.

The Victorian Mansion was two houses together. The back part of the house was the original Chaffin home and the rest of the house was the new addition. The end result was a Victorian Mansion in the grand manner and the latest style.

The readying of the house took every minute of the two weeks. Local people were interviewed for household positions. Help was needed in the kitchen, in the reception rooms and in the laundry. A master gardener with many helpers was required for the grounds.

When the family arrived from New York the household would be in order, and each one would know their duties and how to perform them to the satisfaction of Cutler. If it satisfied Cutler, it would satisfy Mrs. Daly.

Mr. and Mrs. Cutler had two rooms at the top of the stairs. It was ideally located, for all the other household servants were on the third floor and had to pass by the Cutler's rooms on the way to the kitchen. Any questions could be answered and instructions given by Cutler as the work day started.

On viewing their rooms, Mrs. Cutler said, "Mr. Cutler, they are lovely rooms. What a nice view of the mountains on the west side. Let's make it our sitting area and the other room our bedroom. Each room has a closet, and you can keep your clothes in the sitting room. These new closets are better than the standing wardrobes we used to have. The rugs are nice, and will keep the noise down when we are walking around. Mr. and Mrs. Daly do give us nice living quarters. I know ours are better than other servants', for I have talked to the help in the neighboring apartments in New York."

"Yes, my dear, we do have caring employers, and I suppose it is because they both worked hard and appreciate our work and want us to be comfortable in their employment."

"Well, whatever you say, Mr. Cutler, but we are lucky. The wallpaper is so cheerful in the one room, with nice bouquets of flowers in the paper. I will keep all in order for us, Mr. Cutler."

"You always do, and I appreciate your fine work. I couldn't do my job without you. Should we go out onto the grounds and have our tea, Mrs. Cutler? It is four o'clock and I feel tired. The tea and sandwiches will give us a lift and carry us through until dinner. Stena is probably ready for us to join her for tea. I wonder what sweet she has fixed today?"

The Cutlers closed the doors on their quarters and went down the back stairs to the kitchen. There, Stena was waiting with two pots of tea, finger sandwiches, boiled ham and egg salad, and puffs of pastry filled with whipped cream.

Stena remarked, "I was getting very weary waiting for you — you are just in time and the tea is brewed. I put it all on two trays so we can go outside in the garden or on the porch."

"I think the porch, Stena," said Mr. Cutler, "for the sun is in the west now and the porch will shade us. What a beautiful day. I hope the weather will hold until the family arrives Thursday."

"After tea let us walk to the laundry and see if everything is settled there," said Mrs. Cutler.

The threesome sat on the porch at a small round table and took a well earned rest from cleaning, sorting, arranging and preparing. The house would be ready and all well on June 1, 1888. They finished tea and walked to the laundry.

The laundry was a shingled building. The upstairs had twelve bedrooms for men servants of the mansion and grounds. A cookhouse was on the stock farm, adjacent to the mansion. The menservants and gardeners would join the stable men from the stock farm and horse breeding farm and other field workers for their meals. Mrs. Riffle, the wife of one of the cattle wranglers, was the cook for the hired help, while Stena cooked exclusively for the family and their guests. Some said, "Riffle is as good a cook as Stena." Certainly, she cooked abundantly. No one went hungry at a Daly enterprise, for Marcus Daly came from the Ireland of the great potato famine to the "land of plenty". He was a success now, but he had known hunger.

Eddie ate with the other workmen at the stock farm, but

spent time and slept at the mansion. He bunked in the kitchen and started the big wood cook stove every morning. He gathered and fueled the stove all day and banked it at night. Eddie's duties included supplying 500-pound blocks of ice to the pantry coolers. Using a block and tackle he boosted the 500-pound blocks of ice into the upper chamber of the coolers. This was done outside the house. Inside the pantry the ice storage area was covered with oak siding.

Eddie was hired by Cutler in New York to be a handyman. He was just sixteen years old and cold and hungry. He came to the lower kitchen door at 725 Fifth Avenue looking for a handout. Cutler was surprised at himself for hiring Eddie, as he was not a person to act in haste. Eddie was grateful that Cutler would take a chance by hiring him. Eddie did all he could to prove Mr. Cutler's choice was right. Cutler trained Eddie to do menial jobs -- polish silver, copper and brass; fire the stoves and fireplaces, shine shoes, sweep walks, empty the ash and anything else that was needed. Eddie was a devoted, loyal worker. Cutler's judgement had been good, and Eddie was his right-hand man.

Thursday, June 3rd, was the day of arrival. Marcus was so anxious to see his family that he left in the carriage thirty minutes before the train was to arrive. He would rather wait in the lane for them. A platform had been built for passenger unloading and ramps built for animals or produce to board or unload. The train tracks would eventually go to the end of the valley, but then they ended at Riverside.

The time had come; the train was in sight. Marcus' throat filled. Marcus was so proud of his family, and he wanted them to love this country and their new home as he did. He got out of his carriage and stood on the platform as the train came to a stop. He saw Margaret and all four children looking out the window of the Daly car. They smiled and waved madly, then there they were in his arms.

"Oh, Papa, we've had the most exciting train ride and we thought the ride from Missoula here was the prettiest ever," said Molly, and the others chimed in "Oh, yes!" Marcus looked at Margaret and his longing for her almost overcame his reserve. But a proper embrace was in order, and happiness flowed through both of them as they held each other.

"Oh, Marcus, we will have such a good life here with our own. I've missed you terribly!" said Margaret.

"Margaret, me darling, you look wonderful. By heaven, you grow more beautiful every year." He hugged her and laughed, "You look like a bride, not mama to these four sprouts!" He whispered in her ear, "I love you so, you can't know how my love and thoughts are always with you." At that, Marcus called to the children to get into the carriage, and left the matter of luggage to the two servants who had accompanied the family from New York City.

They drove to the front entrance of the house at Riverside, where Cutler opened the front door promptly and, with a broad smile, said, "Welcome to Riverside, Madam. We've anxiously awaited your coming."

The children bolted from the carriage, greeting Cutler as they ran tumbling into the house. "Careful, you children! Marcus II, watch Harriot -- she is going too fast!" said Madge, who was in charge of the younger group. They paid little heed and were happily chattering and "ohing" about their finds.

Margaret said, "Thank you, Mr. Cutler, it seems all is in order as it always is under your care."

Cutler then retired to the butler's pantry to make up sandwiches for the afternoon tea at 4 p.m.

As Margaret walked into the front hall of the Victorian mansion, she exclaimed, "Marcus, it is perfect. Everything was done as the decorators said it would be! This will be a happy place for us, Marcus. Our family will be together at last."

"Margaret, darlin', you know it makes me happy to see you so well pleased. It is because of you and the children that my life and my work are meaningful," said Marcus.

The smallest children and Madge and Molly glanced at the first floor, but were most anxious to see their own rooms. They left their parents in the front hall while they raced up the front stairs to the second floor. As it turned out, the third floor would be the young people's domain. The guests and Mr. and Mrs. Daly would take the second-floor bedrooms.

Marcus and Margaret walked through the first floor. The front parlor was small and intimate, with a piano in the center of the long wall and matching green settees on either side of the fireplace. The fireplace mantle had carved mahogany panels with highly polished spindles around the top of the mirror in the mantelpiece. The fire box was brass with ornate brass andirons. The highly polished mahogany mantle and the brass pieces gleamed in the lamplight, inviting all to come into the friendly parlor.

The children loved the "pouf", but only on special musical evenings did they get into the parlor. The "pouf" was a tufted bouffant upholstered couch that wrapped around a short column in the center of the room.

On the second floor everyone gathered in the family room, a comfortable room with bookcases filled with dozens of books for every age and interest. This was a favorite meeting place after dinner or at tea time for the family to assemble to share the details and plans of their days. A door opened onto a veranda with its spectacular view of the Bitterroot Mountains.

Madge chose the yellow room on the third floor, and Molly, the blue room. They each had their own bath. Marcus II, Harriot and the nurse had two bedrooms with a bath between. Harriot shared her room with her nurse. She was particularly taken with her bed, a charming lacy iron bed made on a scale just right for a little girl who was not a baby anymore. Harriot said, "I love my bed. It is cozy and warm and the sides will keep me safe."

The mansion accommodations were successful. Everyone was pleased with their rooms, and certainly the guests would be happy with their luxury suites. Everything exceeded even Margaret's high standards.

Marcus and Margaret were still downstairs while the children were settling in on the third floor. Marcus said, "Margaret, your room has a grand suite of furniture. I am anxious for you to see it. It impressed me."

"Well, we shall linger no longer, I'm impressed that you're impressed Marcus, for you never comment about the household!"

At the end of the hall on the left was Margaret's room with a burled walnut, hand-carved Russian sleigh bed and a large matching dresser with an ornate mirror attached.

Margaret was speechless. She had seen many grand bedroom suites, but none as beautiful as this. The spread was lavender with lace borders. The two chairs at the end of the room against the window were burgundy. Margaret said, "Burgundy and lavender are my favorite colors and I chose them for the bedroom suite. I knew I'd like it, but I didn't imagine that it would be as wonderful as this!"

The room was like the interior of a jewel box. The oriental rug glowed like a ruby and the lavender silk window swags looked opalescent in the subdued light.

"Oh Marcus, I am delighted. I thank you always for everything. No one ever had a husband so thoughtful, generous and good." She looked into his eyes, smiling.

He gave her a squeeze. "Margaret, me darlin', just to please you and to know you care makes me happy. Come, before we get too sentimental, let us look at my room."

Opening the door just across the hall, Marcus said approvingly, "I like it. It suits me. Thank goodness, me dear, you didn't do the fancy mahogany or walnut. I like the sturdiness of oak, and I won't need to worry about harming the furniture if I sit on it or put my feet on the footstools. We are well suited here. I am happy and thank you for thinking of my preferences, Margaret, for I will never get used to anything fancy. But I do want it for you and the children."

"Heavens, Marcus, the children will be ready for tea and we are still here in the hall. Actually I'm hungry and will enjoy Cutlers' sandwiches and Stena's cakes and cookies. A pot of tea will be just right. You will have time for tea, won't you Marcus?"

"Yes, Margaret, I wouldn't want to miss what the wee ones have to say about their grand summer home. And, I want to tell them of the horses I bought for them to ride. While I'm about it, Margaret, I have a suitable horse for you. She is a grand horse and carries a side saddle and rider gracefully."

"Another surprise, Marcus -- a horse for each of us!" Margaret laughed and threw up her hands, "Please. I can take no more today, It is too wonderful."

Tea time in the children's dining room was cozy and full of excited chatter.

"Oh, Papa, our rooms are lovely and the colors are pretty and I can see forever from my window," said Molly breathlessly.

Madge concurred with Molly, and Marcus II said he wanted to go fishing right away. Harriot, of course, said, "Me too, me too." At that Marcus II made a face, for who wants a little sister tagging along?

When Marcus told his children about their fine new riding horses, they immediately wanted to go to the stables, but it was late afternoon. The barns would be filling with the horses and the busy grooms preparing the day's measures of oats and hay and settling their charges in for the night.

"Not today, my tykes. Tomorrow morning, first thing, your horses will be down on the grounds below the house, ready for you to ride."

"Thank you, Papa," they all said in unison.

"And, Mother," said Madge, "will you go riding with us tomorrow."

"Of course, I have a new horse and saddle, too, and we shall ride every morning we are here. You know how much I love to ride, and I'm anxious to see the horse your father has selected for me."

"I only select the best," said Marcus, "only the best for my family."

Tea was over and it was time to go to their respective rooms and read or play and then bathe for dinner. When they returned from the tea, the maid met Margaret in the hall and said, "Mrs. Daly, I have brought your clothes that were pressed and they are hanging at the front of your closet. I have laid the children's clothes out for them and have instructed the nurse that one of the younger servants should be there at 5:00 to help dress the children for dinner. As for Miss Madge and Miss Molly, I will help them dress and see that all the hooks are fastened and the buttons buttoned."

Margaret said, "Thank you, Josephine. I appreciate your promptness and attention to detail. Tell me, are you from Hamilton?"

"Yes, Madam, I am. My father works as a logger and I help in the rooming house on Second Street during the winter. But I knew I wanted to do summer work here if I could."

"We are pleased to have you, Josephine," Margaret replied.

Each continued on their way, Margaret to her bedroom and a rest before dinner and Josephine to carry out her duties.

At 6 p.m., the children went to the dining room while Marcus and Margaret went to the living room for their cocktails with dinner for them at 7 p.m.

Just like the stately homes in Ireland and England, Riverside adopted the country gentry's routine:

At 8 a.m.: Tea and a wake-up call by the maid

9 a.m.: Breakfast either in the children's dining room or on trays in their respective bedrooms

10 a.m.: Morning activities, hunting, fishing, horseback riding, coaching, etc.

12 noon: Luncheon for those not on all-day excursions (they would have a box lunch). Often luncheon was served on the dining room veranda or on the north porch in the shade on hot days.)

1 p.m.: Rest in respective bedrooms, letter writing or reading

3 p.m.: Tea in the dining room or sun room

4 p.m.: Croquet, bowling on the green, painting

5 p.m.: Dress for dinner in formal wear

6 p.m.: Cocktails served in the living room

6 p.m.: Children in the dining room

7 p.m.: Dinner in the dining room

8 p.m.: Coffee after dinner served in the living room

This schedule was posted in the kitchen, and the guests were informed about the schedule on their arrival. It has been stated that this routine kept the gentry from becoming bored or slovenly or both. A change of dress was required on at least three occasions and little free time left for one to waste or become bored. Books were in cases in the living room, and after the evenings activities most guests chose a book or paper to read later in their rooms. After dinner, some of the men retired to the third floor billiard room for a game, a cigar and brandy.

As the summer went on and the house filled with guests, the evening festivities might include outdoor games. The daylight in the summer was bright until 10 p.m. Then the Bitterroot mountains would be silhouetted against the sky and look one dimensional and unreal, like a cardboard cutout.

The first guests, of course, were Margaret Daly's mother and father, Mr. and Mrs. Evans, and her sister Mattie, and Mattie's companion, Cora. They came by train from Anaconda to Hamilton with a change and wait at Garrison, Montana.

The Evans were originally from Ohio and, like most midwestern people, were conservative — not stingy, but certainly not wasteful -- conservative in their dress, jewelry and everyday living they never boasted, but were retiring and quiet about their business success.

When Mr. and Mrs. Evans, Hattie and Cora arrived at the front entrance and Cutler greeted them, Mrs. Evans asked, "Where is Mrs. Daly, Cutler?" Just as he was about to answer, Margaret came into the reception hall and greeted the family with hugs and said, "I'm so happy to see all of you. It has been too long. Come with me and we will settle in your rooms. Mattie, you and Cora are here on the first floor in the little suite. Mother and Father, come along with me to the second floor. Your rooms are next to ours, Mother, with a lovely view of the grounds and the Sapphire mountains.

Mrs. Evans shook her head in wonderment, "Margaret this is such an extravagance. Surely the house is too big. Poor Marcus! To keep this estate will cost a fortune! But it is beautiful here and certainly this is the grandest home I've ever seen.

"Mother, please do not worry. Marcus does not overextend. This is what he wanted for his family and friends: a place where dear ones can meet to exchange news and to have jolly times together. This will be a peaceful haven for us in this hard world. Just enjoy!"

Both Mr. and Mrs. Evans went to their rooms. All the bedrooms had an identical amount of furniture: a bed, a dresser, a desk and chair, a rocking chair and a chaise. A bathroom was between any two bedrooms. The wallpaper for the Evans' bedrooms was grey with a trellis pattern that was interspersed with red roses. The fabric the spreads and drapes and the chaise were the same. Each bedroom had a personal laundry bag made of fabric matching the fabric of the particular bedroom, enabling the maids to know whose laundry belonged to which room.

The guests and the Daly family all changed their clothes frequently, and the astronomical number of clothes gathered each day astounded the laundry workers. They had large tubs with hard stompers to swish the clothes clean. Sad irons were all in a row on the stove, along with great tubs of water boiling to mix with soap and bring the clothes to extraordinary whiteness. Steam spitting, working noises, chatter and laughter filled the laundry. The laundry house was a good distance from the mansion, and servants were able to release their complaints, gossip and views without disturbing the Daly family and their guests.

Olga was the head laundress and she allowed no foolishness. She often had occasion to call for the girls to settle down. Olga called to Moria, her assistant, "Please, quiet down, and no more complaints about the people in the mansion changing their clothes too often. My glory, if they didn't, you wouldn't have a job!" This was true, but part of the style was to complain, though they looked forward to this summer work. There were not that many jobs available in the valley, and Riverside brought in much summer help. Since it was somewhat isolated from the community the mansion and its support system became a community of its own. The women who toiled in the laundry saw the men workers on their way to work, and many a valley marriage was the result of a summer romance begun at Riverside.

Riverside was a wonderland not only for the Daly family and their friends, but also for the employees. The privacy fence around the servants' entrance to the kitchen allowed them to sit outside and enjoy the view and chat without intruding on the Daly family.

The bell in the butlers pantry was ringing, and Mr. Cutler came from the front hall to answer. The bell box indicated it was Mrs. Daly's room bell. Mr. Cutler answered this call himself, taking the back stairs to the front hall and on to the side hall to the bedroom at the end, which was Mrs. Daly's.

Cutler rapped at her door and said, "Madam, did you call?"

"Yes, Mr. Cutler, please come in. We are expecting our first guests other than family. Dr. Murray and his bride. You remember Dr. Murray from Butte and Anaconda?"

"Yes, ma'am, I do. It will be a pleasure to see him again and to greet his new wife," said Mr. Cutler.

"That will make it eight adults and four children in the house." said Margaret. "The Lewis clan will be coming next week. I think they will have the blue room at the end of the front hall. The Murrays will have the yellow room on the west side of the house and, when the Lewises come, give them the yellow striped room with yellow roses in the drapes. Unless we number the rooms, Cutler, this is the only way I know to designate them."

"That is a fine way, Mrs. Daly, for I know each room very well. I will see that everything is ready. Eddie will meet the train for the luggage, and Martin at the carriage house will drive the buggy and meet the guests on the train at Riverside's lane. Do you want me to tell Stena?"

"Yes, please do, Mr. Cutler. I will talk to her later about the menus, but she could be thinking about them. How do you find the work here, Mr. Cutler? Do you have enough help? It seems to me that all is well, but do tell me if you need extra employees, for we will have a full house before long. Mr. Daly is busy with all the projects on the different farms, so I must take care of the home and garden details with your help."

Mr. Cutler returned to the kitchen with the instructions from Mrs. Daly. Stena was in the servants' dining room reading her Swedish paper. Stena hungered for news of her homeland, and the weekly newspaper was a treat that she cherished.

Cutler said, "Stena, when you are through with the paper, come to the butler's pantry, for I have instructions for you from Mrs. Daly. Mrs. Cutler and I will be arranging the services for tea."

Stena replied, "I'm through now. Mr. Cutler, I've had a good time reading about happenings in Stockholm. I like to read about the ship crossing schedules and when they are due in New York."

Stena was busy in thought after Mr. Cutler told her the Lewises were coming to visit. Mrs. Lewis, Hattie, was a favorite of Stena's, and she wanted to prepare something special for her arrival. The rhubarb was up and she would make some pie, cutting the rhubarb small and adding lots of sugar and butter with a little flour was the secret. Stena would top the pie with her famous crust, then served it warm with double devon cream. Perfect, thought Stena. She would check with the farm for fresh produce and the stock farm for beef, pork and chicken, and plan the menus for Mrs. Daly's approval. Stena called the two young kitchen maids. "Girls, come and help me prepare the menus and then I want you to go to the stock farm for supplies. If they don't have it, tell them to get it."

The stock farm was the main headquarters and adjacent to

the mansion grounds. The supplies for the mansion were dispersed from the warehouse adjacent to the cookhouse. What wasn't grown or ready on the Daly farms, was supplied by local planters or sent from Missoula on the train.

By July, the crops were coming along so that fresh vegetables would soon be on the tables. It was still June and canned fruits and vegetables and the food in the root cellar were being used. Fresh beef, lamb, pork, chicken, and ham were the main courses, with fresh trout as the appetizer. Stena had 101 ways to cook chicken, and she used great imagination in preparing the available foods.

The arrival of the Lewis family was a joyous occasion for again they had not seen the Daly family for nine months. Though Hattie had always thought they would get to New York, Mr. Lewis was busy with railroad business as a director of the Northern Pacific railroad.

After all of the greetings, Hattie met Margaret for a visit in the upstairs sitting room.

The sitting room was conducive to small or large groups and usable by all ages. The bookcases had books for adults, students, and children. There was a toy box in the far corner for Marcus II, Harriot and their young guests. The furniture was soft and heavily upholstered in cut velvet of burgundy and green. A large center table with a paisley fringed scarf divided the room. In the table center was a glass dome with a floral scene. The table was large enough and high enough that little ones could see but not touch. Hattie and Margaret took the two carved Victorian chairs by the fireplace and began to catch up on their lives. Hattie spoke first. "Margaret, New York agrees with you. You're stunning. We do miss you but can't wait to hear about the exciting life you have."

"Hattie, my life is not very exciting. The children require most of my time and thoughts. Marcus is gone so much and doesn't care for social matters, just business. He doesn't like the New York social scene at all and, of course, he is Irish, and the Irish aren't accepted there. The businessmen accept Marcus, but together we do nothing. It is one of the hardships we have to endure because of the children and the educational opportunities needed for them to become responsible and worldly people.

"Marcus is so dynamic that it crushes me to see the shoulders turn as soon as he is introduced. So, Hattie, we have our cross to bear. This snub has caused Marcus to refuse any invitations to social functions. He insists that I go even when he is in town. I'm sorry to burden you with all this. Actually, Riverside was created because we need to be together as a family, Marcus needs relaxation, and the children need to know Montana, their family roots, and the foundation of their father's business and success. Riverside will do that for them, and in our small kingdom here we can be surrounded with family, friends and love of and for them. Our good friends who come to visit provide a solace to the mind and soul for Marcus, and his demanding work is forgotten in his love of the land.

"Enough of this Hattie, tell me about your life."

"It isn't settled yet, Margaret, and you are the first I've told, but we think we will be going to California to live. Now that train travel is coast-to-coast, I won't feel so isolated from family if we do move to California. George has been in San Francisco a great deal and he is sure we will enjoy life there. I have mixed emotions, but it might be for the best. Now that you have Riverside we can be together in the summer wherever we live."

Hattie continued, "Margaret, what about Miriam and J. Ross? Are they coming to Riverside this summer? I see Miriam in Butte and we have our sisterly chats and she always asks if we heard from you, Margaret."

"No, Hattie, W. A. Clark and Marcus are really at opposite polls now. W. A.'s newspaper is publishing editorials that accuse Marcus of wrongdoings. Marcus is furious and intends to establish a paper in Anaconda to counter the attacks. I hate all this. I can't understand how it all came about. It seems to have gone so far that no reconciliation is possible. Mr. Haggin is adamant that W. A. Clark should be humbled. W. A. called Mr. Haggin a 'foreigner from Turkey trying to tell us Americans how to run our country.' Mr. Haggin isn't a foreigner. He was born in Kentucky and his father was of Turkish descent. As far as Marcus is concerned, a slur on Mr. Haggin is a slur on him. You must know that Marcus and Mr. Haggin *are* Anaconda Copper Company. The other partners, Tevis and Hearst, let Haggin make all their business decisions.

"Going to California may be the best for you, Hattie. I should think you would like to be away from all the dissension in Butte. You are right -- Riverside will be a sanctuary," said Margaret.

"How do the children like New York, Margaret?"

"Madge is fifteen and Molly twelve, and they have just finished their first year at Manhattanville School, a very fine girls' school. We have an English nanny for Harriot and Marcus II. Marcus started at a private school this year, and Harriot is trying to read his books! We have a full schedule for the children, with piano for all except Harriot. Even Marcus II is learning to play and, of course, Harriot is begging for lessons as well!

"As to how the older girls like New York, I'm not quite sure. It isn't easy to make friends. One good thing is that the school makes the students feel as if they are privileged to be part of the student body. This gives the girls a bond. I'm proud of Molly and Madge. If they have had a hard adjustment, they haven't told me. I suppose they have each other for support.

"Marcus II is making friends and didn't want to come to Riverside, but he likes to hunt and fish, and Cutler and Eddie take him out to the creek every afternoon. We may have trouble getting him back to New York! Harriot is Harriot, and happy anywhere. She has her own horse now and is learning to ride at four years of age. I think that's too young, but she wants to be able to ride with Marcus. They are very close, you know. He calls her his 'Little Princess'.

"Hattie, do you ever get to Anaconda? Do you see Mary, our dressmaker and shopkeeper? I think Mother does, but I have a personal interest since we brought Mary to Anaconda and helped her set up her shop."

"I see her when I go to Anaconda. I still have her make my clothes. Speaking of clothes, Margaret, what are the latest styles from Paris? We know New York slavishly follows the edict from Paris, and I do like to keep up with the fashions. One of women's favorite pastimes." She laughed.

Margaret was well versed in clothing and told Hattie, "The bustle is still with us. I remember how nice it was in 1882 when clothes had a more natural line, but the bustle is as big if not bigger than ever. I brought a book. Let me get it, and we can look together. Let's move over to the sofa and share."

Margaret went down the south hall to her bedroom and returned with Harper's Bazaar book that had styles from 1886-1888. They looked at ball gowns, afternoon reception frocks, traveling clothes, and riding habits. The mourning clothes they fortunately didn't need.

"Oh, Margaret, look at this reception gown for the hostess of an afternoon reception. It has a train, and is of sapphire blue velvet. The front panel where the velvet panel opens has a cascade of pale pink faille trimmed with blonde lace and gold embroidery. The pink faille front panel is embroidered in gold and draped with lace shirred to the belt and falling jabot fashion to the foot on the left side. The pink silk bodice is trimmed in the same lace and has a dog collar or double pierrot collar of lace pointed low at the throat. The footwear is velvet slippers with pink stockings covering the legs. Oh, Margaret, have you ever seen such a grand dress?"

"Yes, Hattie, New Yorkers dress fashionably and extravagantly. You would find New York society women wearing this hostess gown with pearls, a diamond pin, earrings, rings or bracelets. They display their wealth and seem to be at ease with it all."

"I, for one, cannot stand all that lace and fuss, so I would like the sapphire blue velvet and pink faille but would have less lace and keep the gold embroidery. I do admire the workmanship on the embroidery. At any rate, the style will be everywhere, and Butte and Anaconda will have the patterns from France as soon as possible. Butte and Anaconda society will be entertaining and entertained in similar gowns. Aren't I right, Hattie?"

"Yes, since we have the theater and opera house for the wealthy to patronize in Butte and Anaconda, the latest style is in vogue!" said Hattie with a little laugh.

"Well, let's see if traveling clothes are comfortable. This looks very plain after the hostess reception gown, but I like it." Hattie read aloud, "It says it is a silver grey camel hair dress combined with velvet of a darker shade, an empire sash of grey soft finished silk, wrapped three times around the waist, and the sash ends fall low in front with passementerie tassels at the sash ends. A gray felt hat shaded with a long ostrich plumes is pictured with the costume. Light grey suede gloves and a seal leather handbag with old silver mountings set off the suit. Isn't that gorgeous, Margaret? You should have that."

"Before I left New York I ordered a number of clothes for myself. I have something like that coming, even in grey. Marcus gave me a string of grey pearls that are twenty-four inches long. I'm so proud of them. They are perfectly matched, just like the white pearl necklace he gave me as my first piece of important jewelry. Pearls really are my favorite. I also have a friendship circle diamond pin from Marcus and sometimes I pin both strands together with the pin. As far as having lots of jewelry, I'm really not interested. Our midwestern background doesn't allow me to be so extravagant and I'm not comfortable in large pieces of jewelry. Others wear it gracefully. I'm just not able to do so. It's like a log chain around my neck when I put on the large emerald and diamond chokers or ruby and diamond necklaces. No, I have a nice diamond ring, and an emerald, a ruby and a sapphire dinner ring. What ever made me get off on jewelry? Hattie, it is so wonderful to have you to talk to. I miss you so much in New York, Please always come to Riverside; let's never grow apart."

"Never, Margaret, I will always be there for you and yours. Goodness, it's 3:00. Mattie will be up from her nap, and it is time for tea. It's been fun. Let's run."

Neither Hattie nor Margaret took long in their rooms to freshen for tea as they were late and Mattie would question them. Mattie was so sensitive and would be hurt to know they had looked at *Harper's Bazaar* without her, even though she knew she needed her rest and Margaret and Hattie didn't. Neither Hattie nor Margaret would volunteer their whereabouts the last hours, unless a direct question was asked.

The tea was orange pekoe, brewed in a China teapot from England. This tea set was a very ornate pattern of Crown Derby china called Red Aves, featuring game birds and flowers in a burnt red color on white porcelain. Margaret did not use this for dinner china, as it was too ornate for her taste; but it was great fun for tea, cookies, finger sandwiches of cottage cheese and chives, potted ham, and cookies. There was a large selection of treats but not too much food at tea, for the dinner was to be substantial that evening.

"Mattie, after tea we are going to stroll down the lane. We will walk slowly and take your wheel chair in case you tire. It is such a beautiful day, and the sky is blue, the mountains are dark blue and everything on the ground is spring green."

"Oh yes, Margaret, I would love the walk, and Cora will come with us to push my chair."

They finished their tea and did not linger, as they wanted to be outside on this fine June day in the Bitterroots.

The days came and went with horseback riding, walking on trails, badminton, croquet, fishing, and many happy hours of conversation. There were just not enough hours in the day. It was a wonderful happy summer with activities and friendship.

Marcus was thinking about the thoroughbred horse farm.

He had made extensive inquiries, and now had Sam Lucas as a trainer. This winter he would go to Ireland and purchase the best stock possible. He felt, as many others did, that Ireland had the best racing stock in the world, but he would not limit his purchases to Ireland. Marcus would buy the finest in Europe and the British Isles, as well as in America.

"Margaret, darlin," called Marcus outside her bedroom at an early hour on the 24th of August, 1888. "I'm needing to go to Ireland, and since our summer is almost over I'm going ahead. I'll take the two servants back to New York with me, and you can bring the rest of the staff and the children when you are ready. I also have a meeting with Mr. Haggin in Butte. I want to stop in Anaconda and see about my newspaper project. I've contacted an excellent man about becoming my editor, but I have not heard if he will accept my offer. He should -- it will be a free hand I give to him to make the paper the finest in the West."

"Marcus, you have too many responsibilities and you are pushing yourself too hard." Margaret frowned.

"Oh no, my dear, these few months in the Bitterroot have restored my vigor. I'm ready for anything. The logging here is going well. Mr. Hamilton has finished the dam on Skalkaho for our irrigation system, and I want to get busy with the horses. I intend to make Riverside and each farm, including the thoroughbred farm, a paying proposition. That is the best of it — getting great fun out of your recreation and having it pay for itself!

"I'll be leaving after the weekend and will be back in New York the first week of October. By then, all the children except Harriot will be in school, and your life will be in order, Margaret."

"All right, Marcus, but please get plenty of good rest and

be careful what you eat. You are getting heavier, and I know that is not the best for you."

"All right, me darling, I'll eat soup and vegetables and little bread, but meat, I will not give up."

Marcus left on the train to Missoula in his private car which made connections at Garrison from New York by way of Chicago. His stop in Butte and his meeting with Ben Haggin would take a week, for he wanted to take care of mine matters and the smelter.

On the train to Butte, Marcus was glad the price war with the Michigan producers was over. Michigan had had their way many years with copper producing, and the Michigan group was going to break the Montana copper miners, of which Anaconda was the leader.

Marcus mused that by the time copper price war was over, Anaconda had produced in its one mine one-half as much as the seven Michigan mines, and the other mine owners' production at Butte had powered ahead of Michigan's production, with Anaconda leading them. They had prevailed and broken the back of the Michigan cartel's ability to set the price for the copper industry. Marcus was happy to put that behind him.

Ben Haggin was waiting in Butte to hail Marcus as a man of great vision and ability in winning the battle of the copper pricing. But, this year was not over, and another ordeal for copper pricing had already come about. A French syndicate had been formed by European bankers and other wealthy individuals who envisioned monopolizing the copper market world-wide.

Marcus arrived in Butte and went directly to the mine office to meet Ben Haggin.

As Marcus entered the office, the male secretaries and clerks said, "Good morning, Mr. Daly." The executive secretary for Mr. Daly came forward to meet Marcus and said, "Mr. Haggin is in your office and waiting. I have ordered some coffee, pastries and fruit for your meeting."

"Thank you, Morgan, that will be fine. I'll call you later if we need documentation of our meeting. First, I want to visit unofficially with Mr. Haggin."

As the door opened and Marcus entered the office, Ben rose from his chair and with warmth of friendship greeted Marcus, "Marcus you were right! We whipped those Michigan price-setters! It is great credit to you that you had the foresight and ability to break their monopoly in copper price fixing. You are the talk of the mining world, and we owe you a great debt of gratitude, Marcus."

"Thanks to you, Ali Ben Haggin, Anaconda Copper exists because you believed in an uneducated Irish immigrant. It's a mystery how I can see into a mine and know the potential or how I can plan strategy to preserve what we have. The Irish believe in magic and God. Maybe both have touched me, Ben. May God continue with his hand on me. You also know that we Irish don't get too happy, for that is tempting the fates. We will savor this victory quietly. I've renewed my spirit at Riverside this summer and I'm ready to tackle what comes."

"Good, Marcus, for we have received an offer from a French syndicate to guarantee a copper price of $13 a pound. All the mines here are signing up and I think we should. It will curtail production of copper and keep the price fixed, and there won't be any more price wars."

This was an explosive plan. Marcus turned and looked out

the window at his beloved Anaconda Mine. "Ben," said Marcus, "I don't like it. Anything that curtails production causes unrest. Miners are laid off . . . accusations of greed . . . and God knows what! My hell, Ben, we don't need to be greedy. We're all wealthier than we need to be."

"Sorry, Marcus, but you know I have the votes of Tevis and Hearst. They both have lost interest in the actual operations. Marcus, you do have a free hand to run the company, but we will need to join this syndicate. They are too powerful for even you to handle."

"Well, Ben, I guess I have no choice. Let me call Morgan to write a letter of consent and a contract, but I want it to state that Marcus Daly was not in favor of signing with the copper syndicate."

"All right, Marcus. I'm sorry about forcing you. You know I go along with you 90% of the time," Haggin said apologetically.

"Fine, Ben, I know." Marcus' voice was tense.

Morgan wrote the contract and the consent document to send to the syndicate in Paris.

"Ben, are you going to come to Anaconda with me to see the smelter?"

"No, Marcus, I'm on my way to New York for a meeting with some eastern investors. I'll let you know what is going on there. The moguls in the east are stirring and wish to invest in copper. They invested in cattle and in the sheep industry. Everybody wants part of Montana. We need those Easterners. There is not enough capital here to do all the mining, ranching and developing that Montana offers. We just don't have the population."

"I know, Ben, but take care you don't sell the store."

Ben laughed and said, "Marcus, I'm smarter than that."

Marcus left Ben at the station and boarded the Anaconda train. Marcus felt an uneasiness, but he could do nothing about the syndicate. It was a wait-and-see situation. He thought it could fail.

He thought about the eastern investors who had come to Montana to invest in range cattle and sheep. They came in droves in 1883. The amount of stock on open range tripled, and the absentee cattle owners reaped a profit. But all this ended in 1886-87. Just over a year ago, Marcus mused. They overcrowded the range, the cold winters froze the ground, and then there was a thaw that made ice of the snow. The cattle, sheep and horses could not break the ice to get at the little grass the drought had left. This tragedy broke the open range cattle and sheep industry. One figure stuck out in Marcus' mind: the E6 and Turkey Creek ranches had 27,000 head of cattle; by spring they had only 250 head left alive.

He would have an irrigation system that Jim Hamilton had devised from Skalhako Creek. Even failure from a drought could be avoided. The mountains have snow; it melts; they would harness the runoff and irrigate. It was a horrifying sight to see all the carcasses in the ravines and stream beds after the cattle had starved to death in 1887. The ranchers that survived could no longer turn the herds loose to fend for themselves. They would need to husband the animals. Riverside would have deeded land for the stock to graze with summer pastures in the mountains.

One result of this disaster was that the cattle business reverted to native Montanans. In his heart Marcus knew he was a true Montanan. Riverside would use the latest ranching methods and be a showplace for livestock, thoroughbred race horses and trotters. His farms would have fields of grains, beds of vegetables, hay fields and nursery stock for trees, shrubs and grasses that would be used to prevent erosion of the land. Marcus pledged to himself that Riverside would be an example for all to follow. Riverside enterprises would be innovative in agriculture and husbandry. He had some theories on both, and he would exercise what he believed. The twenty-six mile train ride from Butte to Anaconda passed quickly, and Marcus was called back from his reverie by the conductor.

"Mr. Daly, here is your station. The driver is standing by. I saw him as I looked down the track from the step. A pleasure to see you, Mr. Daly, and I'm happy to be working on the Anaconda Rail line."

The conductor, dressed in his navy blue uniform with his round-billed cap, took Marcus' luggage, set it aside and gave his hand to Marcus as he stepped down. Marcus said, "Thank you, Jim, for taking care of my luggage. I always seem to have more than I need. Tell your family 'hello' and give them a treat from Marcus." The conductor looked into his hand at the $5.00 gold piece. He called after Marcus, "Thank you, Mr. Daly, will do as you say." Marcus and his driver went to the Montana Hotel. It would be a short visit, for Marcus wanted to get on to New York and off to Ireland for horses in January.

Marcus went to the smelter and found Murray, his old friend from mining days who was now the manager. Marcus and Murray greeted each other.

"Thomas, you old leprechaun, how goes it for you? The report from the smelter tells me you are staying ahead of it all. I knew you could do it. You always put yourself down. You've gotten a hold on this, and we're producing copper at record levels.

Keep your men happy, and don't forget them. Their loyalty is what makes the wheels turn."

"Right you are, Marcus. I've learned a lot and I'm proud that you thought enough of my ability to make me your manager. Who would have thought we two Irish lads would be so successful? Course, I'm not as successful as you, but I'm proud to be doing this work. Marcus, you are becoming a real tycoon!"

Marcus laughed, "Well, Thomas, I seem to be driven by ideas and ideals, and I'm fortunate that success is mine. I still feel like that fifteen-year-old Irish boy, and I still go day by day.

"Thomas, I'm off to Ireland. Any messages you want to send to your family? I'm going to some of the thoroughbred farms and it will be no trouble to call on your family in Dublin."

"Well, Marcus, most of my family has left Ireland and are scattered here in the U.S. My sister lives in New York State, but miles from New York City. She is going to come to visit some summer, she says. But thank you for asking."

"I only have one sister left in Ireland, and she won't come to America. She is on the family farm, and I try to help in the improvements. She is stubborn and independent, but she is not starving or wanting, for I won't allow that," said Marcus. "Let's go check the smelter, Thomas. Show me what you've got!"

The two men started through the smelter, with its belching smoke and spitting fire. The raw ore was melted and poured into molds to be made into bars, cakes and ingots, which in turn were to be made into useful products such as copper sheets, wire and pipe combined with other materials to form alloys. The noise of the furnaces and machines made it impossible to converse, but they could see the process taking place.

The huge smoke stack outside was a landmark for Anaconda and could be seen from miles away. The smoke belched out into the rarified atmosphere, leaving clean air at the level of the town.

Within the week Marcus took the commuter train to Butte to connect with the main line to Chicago and New York.

Arriving in New York around 4:00 in the afternoon, Marcus took a Handsome cab to his home 725 Fifth Avenue. The two servants that had accompanied him from Riverside rode in a carriage behind and took care of the luggage.

It had been four weeks since he had left Margaret and family at Riverside, and they had arrived in New York a week earlier. As he went into the lower level and was greeted by Cutler, he asked, "Where are Mrs. Daly and the children?"

Cutler said, "Mrs. Daly is receiving guests in the front parlor and Miss Madge and Miss Molly are having piano and singing lessons on the third floor. The smallest children are at the park with the nanny."

Marcus was disappointed that no one was there to greet him, but he went into his suite of rooms on the third floor, hoping the noise from the music room would be muffled enough that he could rest and have a highball. He instructed Cutler to bring him a tray with some Irish whiskey and a bowl of peanuts. Cutler said, "I will bring them immediately, I don't think Madam knew you would be here today."

Marcus said, "Probably not. I didn't know, myself. I left Anaconda to come home when my work was done."

Cutler noticed a little irritation in Mr. Daly's voice. "Mr. Daly is tired," he thought.

This same scene met Marcus many times while he was in New York. Everyone busy with their own activities, causing Marcus to seem like an intruder. He knew he overreacted. They couldn't drop everything to greet him. Still, this wasn't home to him.

Marcus was leaving for Ireland in January and asked Margaret if she would like to come with the children. Margaret replied, "Marcus, dear, you know the older girls are going to be enrolled in school in Paris this semester. Harriot and Marcus II will be at home in Paris and will be tutored in their school work. I will need to be there, but you will be coming to the continent after Ireland, so that will be our time together. I just can't find the time to go to Ireland."

Marcus knew that Margaret was doing what she thought best for the children, but he was hurt that his wishes and well-being were not considered. Then Marcus thought, "how silly of me to be so sensitive," and he threw himself into the thoroughbred horse buying business. The horses were running at Belmont Park in New York for two weeks in the fall and again in the spring for two weeks. Marcus went to the races in November and bought the winning horses whenever possible. He did this all over the U.S. and Britain either in person or through a network of agents who acted for him.

Christmas came and there were great festivities for the children and many galas for Margaret and her growing social circle. Marcus was very lonely. This New York high life at 725 Fifth Avenue held nothing for him.

The day after Christmas 1889 Marcus went to the Nederland Hotel which was close to the Fifth Avenue apartment, and engaged a suite for himself. At least he could have a place for himself and could have friends or business acquaintances call there as well. The Nederland had a three-room suite, a bedroom and bath, a living room, a dining area and a kitchen. There were servants rooms at a lower level if he wished to bring servants of his own, but he didn't need a servant because he intended to use the hotel staff for service. When he didn't join the family at 725 for meals, he would have food from the hotel kitchen or eat in the hotel dining room.

Marcus said to himself, "By gad, I feel better. That apartment smothers me with all the French furniture and glitter. Margaret will be able to entertain the social set without having to explain my presence and apologize for my Irish heritage. The children can say, 'Papa is busy with business. When I do see them, they will be with me for a visit and not running off to see or have friends at home. This will be better."

"Margaret," Marcus called as he entered her sitting room, having checked with Cutler that she had no visitors, "where are you?"

"Marcus, is that you? I'm coming. I just finished dressing for tea. Are you going to have tea with me?"

"Margaret, I need to tell you what I've done and, no, I'm not having tea. I'll have some Irish whiskey."

"What is it? I can't tell if you are irritated with me or anxious to tell me what you've done."

"Both, me dear, for I practically need an appointment to see you. I've decided I'm suffocated in this apartment. I am not fond of the social life that you have, which is fine for you and the children; I've taken a suite at the Nederland Hotel; and I'm quite pleased with it all."

"Marcus, how could you, what will people say? Marcus, are you leaving me?" This news had completely shattered the calm of Margaret Daly. This threw her into hysteria. "What will people think of us?" She wept.

"Margaret, hush, I'll be here whenever you want, but when you are busy with social obligations and the children's social life as well as school, I won't be in the way. I don't care about what people say, but I do care about you and the children."

"Marcus, you are never in the way. How could you say this to me? How can you disgrace me? You can't mean it."

"Margaret, when your friends ask, you will say I'm away on business; and if later they find out I have a suite at the Nederland, you can tell them I have it for business reasons. All right? Now stop crying and carrying on. You're still the light of my life and reason for being. You and the children are everything; and don't think I've given up my marriage bed, me darlin'. I intend to visit you often."

"Marcus, I can't believe what you've said. You haven't given me time to absorb this shock. We have to leave for France in a week. Oh, Marcus, what are you doing?"

Thus life in New York changed for the Daly family. Marcus resided at the Nederland and visited his family at 725 Fifth Avenue. Margaret adjusted to the arrangement and, since the children still saw their father at dinner or tea time, life didn't change that much for them.

Marcus was happy with the arrangement and since it was to be, Margaret adjusted her life so that Marcus had time with the family without interruptions. The social demands were met by

Margaret. Even she had to admit that Marcus made her very nervous when her friends called. Marcus had no use for the endless teas, seven-course dinners and the string quartets. Their entertainment reminded him of a funeral dirge. Unusual as the arrangement was, it worked for them.

Chapter Six

On January 10, 1890, Marcus went to the apartment to bid goodbye to Margaret and the children. As he entered the foyer of the building and was about to get aboard the glass and gold elevator, Madge and Molly came through the entrance. "Papa, wait for us. We've been to the park, and we're freezing. The sun was out when we started, but it began to snow while we were in the park. Mrs. Cutler is coming behind us. Oh, there she is. Mrs. Cutler, we're going with Papa on the elevator," said Molly.

"All right, young ladies, I will take care of your boots in the entrance here. That was a brisk walk. Next time we had better look to the east before we start into that park. It's a long way back from the center of Central Park."

"But it was fun, Mrs. Cutler and we are just frosty, and not too wet," said Madge.

"Hope you are all right, Mrs. Cutler," said Molly.

"I'm fine, Miss Molly. Stena always has a hot stove in the kitchen and I'm going there first to warm myself and dry out," said Mrs. Cutler.

Marcus held the elevator while his two oldest daughters entered.

"I'll operate the elevator," said Molly officiously. "I feel so important pushing the lever to close the door and the lever to start. Isn't this amazing? And it certainly would be hard to climb all those stairs, wouldn't it? Which floor, Papa?"

Marcus smiled, "I want to see your mother first. Then, we'll all have tea together. So, the second floor, please, Miss."

"All right, Papa, Madge and I will go up to third and get into some dry clothes. Then, we'll come down for tea and treats, and I guess to say goodbye, at least for a while. Mother says you'll join us in France." Molly's eyes were sad and Madge ducked her chin to hide the tears that threatened to spill.

"Actually, girls, I wanted you all to come to Ireland with me. You need to be acquainted with your heritage. But, your Mother has you enrolled at Madam Coteries school in Paris," Marcus said, and there was a wistful tone in his voice.

The girls were aware of this, but they were so anxious about their own plans: Paris, school, sightseeing and boarding ship, that they soon forgot their Papa's plans and wishes.

Marcus entered the sitting room on the second floor where Margaret was sitting at a desk making lists. Margaret made list after list after list, Marcus thought, and it amused him that she did so. Margaret was concerned with proprieties and she spent a great deal of time writing down procedures governing relationships with servants, as well as for meeting new people, treatment of sales people, shopping and protocol for formal dinner, and other fetes. Marcus said, "Margaret, me darlin', there you are at your lists. Surely by now you know all procedures by memory."

"Oh, Marcus, there you are! These lists are for packing and our departure. Don't make fun of me, it's the only way I can remember all these details."

"Not for a minute, darlin', am I making fun. I just have to tease you a little so that you relax a wee bit."

"Marcus, Marcus, why don't you come on the same ship with us to Paris? And then you can go on to Ireland."

"I have made definite arrangements in Ireland to go to some of the breeding farms, but I will get to Paris in a month or so." He smiled.

"I have sublet an apartment in Paris and the girls can come from boarding school on weekends. It will be more comfortable for them in a foreign country to know that I am close by. Marcus II and Harriot will be tutored in French by an English woman. The French never speak English without the heaviest French accent, and even if the English speak French with an English accent, I think our children will be able to learn the language more quickly with an English person teaching."

"You know best, Margaret, and I suppose if they are to spend time each year in Paris they will need to speak French. I do want you to think about a trip to Ireland on one of these Atlantic voyages. It is their heritage."

"Oh Marcus, Ireland is in such a turmoil, and I really don't think it helps for the children to know all the trials of the Irish. Besides, most of your family, except Honoria, is in America now, and we never see them."

Marcus decided to leave that alone, for he knew why his family didn't visit often. They felt uncomfortable with Margaret's formal ways and, like Marcus, they avoided socializing except with family. Marcus admitted this hurt somewhat, but he did understand Margaret's efforts to offer the children a place of respect in the world of the wealthy. And, even though Marcus never felt wealthy nor took time to count, he did know his children would need to be schooled and educated. He did not want them to work and toil as he had, nor would they need to. If the family took care of their inheritance and built on it, that would be a life's work. Marcus knew that to prepare the children for their place in the world was Margaret's greatest desire and most careful plan.

"Margaret, I'm leaving Saturday for Ireland. I'm going to

sail to Galway and then go to the town of Cong where Ashford Castle is located. The Guinness family have invited me to spend a week with them at the castle and look at some of the horse farms in the area."

"Marcus, I didn't know you knew the Guinness family. Aren't they the brewery people in Ireland?"

"Yes, they are. I've met some of the men here in New York and since I'm Irish and have done well in America, they extended their hospitality to me. Funny, in Ireland they consider me a native son of Ireland; but in America I'm Irish."

"I'm going to have tea with you and the children, then I'll be going to the Nederland Hotel for a business dinner, and tomorrow I will be packing for Ireland. This is our last visit until we meet in Paris. Be careful, my love, take care and I will be there in a month. Do you have all you need, Margaret? I set up a Credit of Exchange between our New York Bank and the Rothchild's Bank in Paris. There is no limit on it, Margaret, nor do I worry. You are a careful manager."

"Marcus, I will miss you. You know how I hate our being apart. It won't be forever, Marcus. Our children will grow up and be gone before we know and then we can have a more normal life together. I will go where you go and whenever."

"Right you are, my dear. You are a wise one, too. I do love ye, Margaret."

At that, Mrs. Cutler entered the room and said, "Tea is served with the children on the second floor. They are already there waiting on you, Sir and Madam."

The tea room was buzzing with conversation as Marcus and Margaret entered the room. The children leapt to their feet and

Marcus II said, "Good afternoon, Mama and Papa. Come -- we want to try Stena's new cinnamon rolls and lemon cakes."

"All right, my loves," said Marcus, "Let's begin. First your Mother will choose. Then we will pass the food around the table."

"Madge and Molly, please help Marcus II and Harriot before they destroy the sweets," said Margaret.

The tea ceremony began with warm milk for the children with a touch of chocolate for flavor, tea for Margaret, and for Marcus, coffee with a jigger of Irish whiskey and topped with heavy cream.

The cinnamon rolls were warm with butter running into the rounds. The sandwiches were chicken breast with mayonnaise, lettuce, thin sliced roast beef and horseradish sauce lathered on brown bread. Slices of cheese, green olives and pickles were laid in a floral pattern on a tray and garnished with fresh green parsley, and the lemon cakes had a dollop of lemon curd on top laced over with sweet whipped cream. The children politely took the sandwiches and relish, but were not as refined in choosing the lemon cakes. Margaret said, "Harriot, don't grab; and Marcus II, don't push. You'll have your cake. There is plenty."

After they had all finished visiting and eating, Marcus said, "Children, I am leaving Saturday for Ireland, and you will be leaving for France soon on a different ship. My plans are to be in Paris in a month to see you, and I hope that you all will be well and happy and will work and enjoy studies in France. You know I will miss you, but I will be very busy looking at race horses not only in Ireland but also in England, France, and Germany. I'm looking forward to my buying trip and plan to take back to Riverside the finest thoroughbreds offered today. Take care of your mother and be sure to keep your diaries so you can remember all about being aboard ship. I'll want to hear of your adventures. Bye, me darlings. Papa loves you."

Each of the children came to their father solemnly and kissed his cheek. Harriot's lip trembled as she fought back her tears. Marcus gave her a big hug and grinned, "Now, my ladies, do not weep when papa must go away for a bit."

The children trooped off to their rooms and Margaret gave her cheek to Marcus for a farewell kiss, but he turned and kissed her full on the mouth.

"Marcus, the children!" Margaret said.

Marcus laughed and said, "Margaret my love, they will have to know sometime that you and I have a private life apart from the one we share with them."

On 10 January 1890 Marcus's ship departed for Ireland. He had been so busy with business arrangements and family affairs he had not thought very much about the journey.

Aboard the ship Marcus settled into his stateroom quickly and decided to go on deck into one of the lounges and watch the New York skyline vanish in the distance.

As he settled into the bar area and ordered an Irish whiskey, he began to think of the day he boarded the sailing ship in Dublin as a young boy. Actually, reminiscing was not something in which Marcus indulged very often, but with nothing pressing him for his attention and his business matters left behind in capable hands, he remembered vividly his leave-taking of Ireland.

Neither his ancestry nor his lineage was of interest to him. As far as Marcus was concerned, he was an American, and life for him had begun when he came to America. What he had accomplished there could not have been done elsewhere -- certainly never in Ireland, which was still a poor and oppressed country.

Not wanting to linger on the depressing conditions of his early life, he moved from the window seat to the bar itself. Others had gathered there, and Marcus was soon engaged in conversations with a young man named John Harrison, and his father, Tom. In men's conversations always there is an introduction of life's work. Both the father and son, John were geologists. This interested Marcus since his own geological genius was from instinct and practical experience rather than from education.

Marcus told the Harrisons he was with Anaconda Copper and that some of the young men in Montana who were sponsored by Anaconda Copper had begun studying geology in college.

When the dinner gong rang and the steward announced dinner would be served in thirty minutes, the men were startled that they had been so immersed in their conversation that a good deal of time had slipped away. They laughed and said they would meet after dinner for a walk around the deck in the brisk air.

Marcus returned to his stateroom and realized he would not be able to attend the captain's Bon Voyage party. He would sit at the Captain's table and make his apologies for his absence at the party. It was not a big issue, but Marcus was a man of his word; he didn't like failing to honor his commitments and he had planned to attend the captain's gala.

Marcus hurriedly changed into formal dress and went immediately to the dining room. He gave his invitation to dine at the captain's table to the steward who was stationed at the entrance of the first-class dining room. The captain was not there yet, for he was first and foremost a seaman and wanted to check the ship and know that it was on open sea and secure. Only then would the captain be able to enjoy dinner with his passengers.

Marcus was seated next to Mrs. Harrison, and Tom and John were seated on the opposite side with Mrs. Cowan between them.

"We meet again, Mr. Daly, and it is a pleasure. This is my wife, Mrs. Harrison, and I've already given her some background about you. We didn't realize we would meet here at the captain's table. Wonderful! It will make our journey interesting and stimulating to have you as a dinner companion."

Marcus replied, "Mrs. Harrison, it is a pleasure to meet you and an honor that you are a dinner partner. Your husband and son and I had a spirited talk about our respective life's work. They go hand in hand: first comes the geological studies, then comes the production of the mine as a result of the geological findings."

The captain came through the door and made his way to the table. He bowed to everyone, greeted some of them by name and then was seated.

The waiters started the process of serving the champagne and the first course of the seven-course dinner. Aboard ship, food was a major diversion, and the finest chefs worked with the finest ingredients to present the diners with menus that would have been appropriate in a five-star restaurant.

The captain asked the guests at his table to introduce themselves and give a bit of information about their backgrounds and locations in the states or abroad. The majority were from New York and other east coast areas. When Marcus said he was from Montana, one lady turned to her husband and said, "Where on earth is that?" Marcus loved to do this. The east coast people really did not know that there was anything between New York and the California coast. In between was of no interest to them. Of course, some of those involved in transportation and investments knew of Montana, its rich copper mines, and the enormous smelter of Anaconda Copper Mining Company.

The dinner proceeded with interesting conversations among the guests and the Captain occasionally telling a story of the sea and ships. There was a lavish display of food and talent. As the first course of poached fish with hollandaise sauce and cheese rolls was served, Mrs. Harrison said to Marcus, "This looks especially good as we get so little seafood in Denver. Actually, as a youngster I thought the only fresh fish were Colorado stream trout and halibut."

"This is delicious," said Marcus. "I'm a great lover of food, but I should cut back as I'm rounding out. My wife, Margaret, is warning me that too much of anything is not good for my health."

The courses came one after the other along with the appropriate wines. The first course had a Chardonnay to complement the flavor of the fish and Hollandaise sauce. The salad was Nicoise, chopped onion, celery, carrots, peas, cubed potatoes, and black beans tossed in a vinaigrette dressing, served on a bed of red leaf lettuce and laced with anchovies on top. Everyone happily engaged in conversation and continued eating and sipping wine as the courses kept coming. The main course was presented by a waiter carrying the silver tray first to the captain and then around the table for the guests to see the chef's perfect offering. The beef tenderloins were whole, browned and crisp outside with rosemary crushed on top. Aside the tray were two silver sauce boats one with brown sauce with sherry, the other with au jus. Surrounding the filets were parsley, potatoes and plum tomatoes rolled in bread crumbs and baked in butter.

When everyone had "oohed" and "awed" and showed their appreciation, the tray was returned to the serving cart and, with a flourish, the head waiter carved the tenderloin with the juices bursting from the first piercing of the meat cooked to perfection from outside, which was well done, to medium and then to rare in the center. All were interested in the procedure, but as the main course was being served and the burgundy wine served, the conversations resumed.

Tom Harrison spoke across the table to Marcus. "Marcus, how long are you staying in Ireland? We will be in Ireland in a month. We're going first to England."

"It shouldn't be longer than two weeks. I can't stay very long any one place, for I need to get my horses purchased and ready to be shipped to America," said Marcus.

"That's too bad,"said Tom. "We'd like to spend more time with you, and Ireland seemed as if it would be the ideal place."

"'Tis a shame we can't meet in Ireland, but we will meet either in Montana, Wyoming or Colorado. We'll keep in touch.

"My gad, I've seen more food served here than anywhere, and I've seen a lot of food in my time," said Marcus.

At that the waiter removed the main course plates, and a small sorbet to cleanse the palate was served in crystal compotes. Everyone enjoyed the cool and refreshing lime sorbet and were ready for dessert and coffee to be served.

The pastry cart had a selection of sweets: chocolate mousse, a Bavarian six-layer confection, chocolate cake, cream puffs, blonde and dark chocolate eclairs, cherry torte and a cheese tray with fresh fruit.

This meal took an hour itself, and with conversation and a rest between the courses the two hours before the entertainment at 9 p.m. passed quickly.

This abundance of food and wine was served at all three main meals in addition to afternoon tea and a midnight buffet. After dinner, Marcus and Tom went for a walk around the deck, and John and his mother went to hear the musicians in the main reception room.

Marcus was to land at Galway where he was to be met by someone from Ashford Castle. Marcus was to be a guest of the Guinness family. This was the first time Marcus had been in Ireland since he left as a fifteen-year-old. So much had happened in his life that Ireland had been pushed back in his mind. Now, with his arrival after so many years away, the memories came flooding back and Marcus was nearly moved to tears. He didn't imagine that he would have this emotional longing and sentimentality about Ireland, and was at a loss to explain the tears that were welling up in his eyes. Was it because the hard shell he had built around himself was punctured by his aloneness and the time he'd had for thought while on this voyage? Marcus was not going to be maudlin. He decided to look forward to seeing his homeland and to returning as a successful businessman. Actually, inside, Marcus still felt like that fifteen-year-old lad with more dreams than experience. How could that be, when he had lived such a demanding life? Some people plot and plan their success and work to succeed. Marcus just worked and planned and was grateful and amazed at his success -- that is, when he took time to think about it at all. Action was to Marcus' liking so ten days on this liner had him ready to jump ship. He started walking the decks for longer periods each day and strained his eyes the last day at sea, looking for his land of Ireland.

Then suddenly there it was, Galway Bay with the Aran Island across its entrance and an armada of other small islands tossed here and there. The land was a green jewel with a slight haze on it. No doubt, it was drizzling rain. All four prevailing winds brought rain, and fifty inches of moisture fell each year on Ireland. The warm gulf stream made Ireland balmy, and the green countryside of Ireland was there year-round.

The ship slid through Galway Bay into the port. The crews were busy with the gangplank, commercial goods, livestock and the passengers. Porters were scurrying around to get luggage, unload trunks and assist travelers ready to disembark.

Marcus thought, I don't care how much you see of the world, it is an exciting moment when you reach your destination and are ready to follow the next adventure. He wished Margaret were here and knew, she would have liked the sights of the harbor in Galway and Ireland. Everyone does.

"If only the political climate in Ireland was not such a holy mess," Marcus thought. The Irish are rebelling against English rule and Marcus had been approached for funds. What could he say or do? Marcus would be subject to some of this harassment in this journey to Ireland, but he was there for one reason — horses -- and he would not be drawn into this political fever. He would try to help the Irish though legitimate channels by using the United States diplomatic corps. Marcus had appealed through the corps to try to influence America to put pressure on the English to stop oppressing the Irish. Marcus had some influence politically and hoped the Congress would support the Irish. At that time the English were omnipotent and world leaders. The English would listen to no one. Marcus had met with little success, but he had not given up. "Surely the U.S. should have some influence in helping Ireland," he mused. Secretly, he thought the U.S. bowed to English decrees just as a child would to his parents.

Marcus was on deck but protected from the wind. The crossing had been very good, unlike that first crossing when the sea had heaved and the ship had bucked like a Montana yearling. Tom Harrison came up behind Marcus and said, "How does it feel to be coming back to your homeland after these many years?"

"I consider America my home, but I will admit I'm excited. I remember the land, the smell and the people well. Of course, many Irish are in the mines and the smelter in Montana, so the people have always been with me. But I had forgotten the land and how precious it is. Even though I've been gone thirty-two years, the die was cast with the Irish brand when I was born. It is wonderful to be here and I'm anxious to travel the countryside and see if the horseflesh is still the best."

"We will be parting in Galway, Tom, but I've enjoyed meeting you and your wife and your son and I am interested in your geologizing of Wyoming's northern section. I intend to keep in touch with you," said Marcus.

"Nothing would please me more, Marcus. I'm living in Denver, Colorado, but the majority of my work is in Wyoming. The family likes Denver's climate and the size of the city better than the harsher climate and isolation of Wyoming," said Tom.

The men parted ways and went to their staterooms to see if all their luggage was ready to be unloaded. Since it was gone already, Marcus checked the room and went down to the lounge to wait to disembark.

The Captain was busy on the bridge, but earlier in the day he had bid Marcus farewell and hoped he would be aboard on his return to America. Marcus didn't bother to explain he would be on a different line on his return. For after Ireland, he would travel Europe on his search for the finest racing thoroughbreds, ending in France to be with Margaret and the children.

The town of Galway was a busy seaport and the safest port in Ireland. The town was irregularly built and the See of a Roman Catholic bishop. Galway was a county by itself. It was at the mouth of the river Corril, on the north shore of Galway Bay. It had extensive salmon and sea fishing and exported agricultural produce, wool and black marble. Marcus observed all of this from the deck, and as the ship docked, the dock hands and cranes went into action and the gangplank was released and put in place. Below, people were crowding the fence which kept them from the passengers until the passport and customs inspections had been completed. This was a long session and, Marcus thought, the slowest unloading of luggage ever. He thought he had waited unreasonably long on trains, but this was worse, much worse.

Finally, the luggage was in place. Marcus claimed his and found the customs man to take care of his belongings. He passed through the passport check, then heard his name called as he exited the passport aisle.

"Mr. Daly, Mr. Marcus Daly -- here, sir." Marcus turned to see a young man in morning clothes coming towards him. "Mr.

Daly, sir, I am here to take you to Ashford Castle. Mr. Guinness is waiting to greet you. I am his personal secretary. My name is Michael Tyrone."

"I'm pleased to meet you. I was wondering how I would make arrangements to get to the town of Cong and Ashford."

"Mr. Daly, Mr. Guinness would never let you be on your own. You are his guest. Our carriage is here, your luggage is being taken care of and we should be on our way shortly. I trust your passage was good. I understand the weather was unusually good."

"You are right, Michael, the crossing was not rough, which is unusual on the Atlantic northern route."

The ride to Cong was four hours and it was necessary to stop for tea at a pub. Of course, it was a Guinness pub, for Guinness beer and pubs were almost a monopoly in Ireland. From Galway to the first stop, Marcus was aware of the sea and the fingers of land intruding into the water. When they left the coast-line, they saw the white heavy-coated sheep in the pastures of green. Never a weed in Ireland, it all looked as if the land had been groomed with scissors. Marcus thought of the story of St. Patrick, an Englishman. St. Patrick was not a myth; rather, he came to Ireland as a priest and rid the island of snakes ("Maybe," thought Marcus, "weeds, too.") There were many sacred places where St. Patrick had been. The age of scholars and enlightenment at that time had been the golden age of Ireland. Irish priestly scholars with brown robes and shaved potes, hair long on the sides and in the back, had taught in England and, indeed, all over Europe, even influencing King Charlemagne of France.

Later in history came the English rule. England depended on Ireland for produce for their people. It came to be that the

English thought that they, as managers, could be more productive on the Irish lands than the Irish were, so they bought or confiscated the land. The English became the landowners, and the native Irish became the poor worker or "crofter" (share cropper). It was a sad time for Ireland, and it was still in a terrible state under the English in 1889.

Marcus was drawn back from his thoughts by Michael Tyrone asking, "Mr. Daly, tell me something about Montana. I can't imagine what it would be like."

Marcus replied, "Well, Michael, Montana is big and varied. Where I have my Riverside farms it isn't so different from Ireland. We don't get the rainfall, but we dam rivers and streams and artificially water by irrigating the fields. We can't get as green as Ireland, but in the spring sometimes the light on the fields and foothills are as green as the Irish green. This is what drew me to the Bitterroot Valley in Montana, for it was there I felt a touch of Ireland in the land.

"There are mountains bigger than here in Ireland, and very rugged and wild. There are bears, cougars, deer, elk, moose, and lakes and streams full of fish. Riverside is in a beautiful valley between two mountain ranges, the Bitterroots and the Sapphires.

"Montana is a wealthy state in resources, and copper is the richest of the resources. The copper mines of my company, Anaconda, are in Butte, Montana and we are mining heavily there. The town is the largest in the state in population. We have Irish, Scandinavians and some Chinese -- a real mixture in background, but they all turn into Americans.

"The country around Butte is bleak, but not far away are beautiful mountain ranges with valleys and snow-capped peaks.

"Actually, Michael, I never thought about it before, but Ireland is like a mountaintop with its lakes, rivers, rounded mountains, and misty rain coming down almost daily. Only Ireland has a warmer climate and a kinder one. You know, you don't have climatic catastrophes such as hurricanes, earthquakes or blizzards, and it seldom freezes or snows. It's the warmest land of all Europe. No wonder they say 'a touch of heaven is Ireland.' The potato famine was its worst disaster, and that was man-made from greed. Even the plagues that wiped out huge populations in Europe never came to Ireland. It's amazing when you think of it. The only thing people die of in Ireland is old age or the one-time famine. No one uses a doctor here, for even flu is rare.

"Here I was telling you about Montana, and I got to comparing Montana to Ireland. You would probably like Montana, Michael. A man can be anyone or do anything he pleases. It is a free state in a free country."

Marcus continued talking until he noticed some men digging in a field close to the road.

"Look. Michael, I almost forgot about digging up the peat for the fires. How could I forget that? Seems strange that you can dig up what looks like dirt and it will burn. A good fire it makes and a warm one." The men dug squares of earth and piled them on a cart drawn by a donkey. They had dug an area about three feet and eight feet wide, which filled their cart, and were ready to leave. The peat is unending and has served as a source for heating these many years of Ireland's existence.

Just ahead was the pub and an inn where they would have an afternoon meal before carrying on to the town of Cong in county Mayo. The driver drove the team of horses up to the entrance of the Fox Inn. As Marcus and Michael were helped out by the driver, the door opened at the inn, with the keeper inviting them to enter. The pub was filled with men having their pint of stout after work, for it was 4:00 in the afternoon. The conversation came to a low murmur as Marcus and Michael entered.

Marcus said, "Hello me fellow Irish, I'm being bold and asking ye to welcome me to my first visit to Ireland since leaving this land to make my way in America. I'm asking the barkeep to set up a round of drinks for ye all and we will toast the Irish and Ireland."

A roar of approval went through the room and all were in favor of welcoming Marcus to his homeland. It seemed the room warmed and glowed in the fellowship that followed. This was Marcus' first welcome as a native son. There would be more welcomes but none that suited Marcus better. Marcus understood the working man and felt a kinship with him, for he, too, had been a laborer. Most men who had wealth such as Marcus did not rise though the work force to their position of power. These men were difficult for Marcus to relate to except in business discussions.

After a short time the driver entered the pub and went to Michael to tell him they must be on their way if they wished to reach Ashford Castle by 6 p.m. It was not far to Cong and Ashford Castle but the horses could only go a certain pace. The carriage was waiting, and Michael and Marcus settled in the carriage and pulled the robe around their legs for the evening coolness was descending. Cong was ahead, up the hill. As the road flattened out, Marcus saw the small town with its narrow street, stucco row houses and shop doors directly off the cobble stone streets' side-walks. The doors were down on the shops, for they had closed at five o'clock. The carriage came to a town square, which was more like a circle. In the center was a large Celtic cross carved from stone.

Ireland was settled by the Celts, and although all of Europe and the known world had been invaded by the Romans and ruins of the Roman civilization are in evidence everywhere, the Romans did not get to Ireland. Maybe they thought the island was not worth the struggle. As a result, the Celtic culture was the background and the backbone of the Irish.

Marcus noted as they left the town that the road narrowed and the closer they got to Ashford the more narrow the road became. The River Cong was beside the road, and they seemed to be following its path. The trees parted in front of Marcus, exposing the splendor of Ashford. The failing light made the castle an outline against the sky, but the grounds were ablaze with torches. With the torches and the little sunlight left, Marcus could see it was a silver castle. Everyone has seen brown castles, but this castle was silver. It almost had sparkles in the stone. It was grey granite. The castle was a fortress, for in bygone days there had been raiding neighbors. The towers had slots through which the bowmen shot their arrows from crossbows. At one time there had been a moat, but it had been filled and became a circle drive, which would take the carriage to the front entrance of the castle. The older part of the castle was used as public rooms and for the household help. The newer part was in the style of a country manor house and, although it was joined together and architecturally pleasing, one could see the difference between the medieval tower, the feudal fortress castle and the renaissance style of the new addition with a Tudor-style section in the center portion. The circle drive was surrounded by grass, and the center of the area had a pond with a fountain. Vines grew up randomly over the castle, partially obscuring some of the windows.

Marcus thought it was all magnificent. The river Cong was making its way by the castle, exiting from Lough Corrib, the large lake behind the castle. Marcus got a glimpse of the lake as the carriage passed the edge of the castle onto the drive and the front entrance.

At the entrance, the door opened and a butler came to the edge of the steps as the footman opened the carriage door. Marcus stepped down with Michael Tyrone just behind. The butler bowed and said, "Mr. Daly, welcome to Ashford Castle. Mr. Edward Guinness is waiting for you in the drawing room."

They entered the huge chamber and Edward Guinness came forward to clasp Marcus' hand and said, "Marcus Daly we are so pleased to have you, our native son, return for a visit to your homeland. We have been aware of your career with Anaconda Copper and are unaware of the years in between. We will have time to reminisce about your life and mine. I enjoyed our last meeting in New York. Since then my father and founder of our family business has died. He was a great man, and I miss him every day.

"Enough, Marcus, come in. Our butler, Mangus, will show you your rooms. We will serve dinner at 9 p.m., a little later tonight, because of your long journey. Will that be enough time for you to change and join us here at 8:30?"

Marcus replied, "You are so kind, Mr. Guinness, and I appreciate your concern. Yes, I will be ready to join you by 8:30. I unpack and pack so frequently it doesn't take me long to settle in. I must tell you how impressive Ashford Castle is. I've never seen a silver castle before."

Edward replied, "It was constructed of granite from a nearby quarry. It is not big deposit, but it was used on the original castle, so we've used it on the newer part."

After the introductions were over, Mangus showed Marcus to his room. His room faced Lough Corrib, and the view from the window was spectacular. In the distance were the mountains to the southeast. Islands dotted the lake close to the castle and Lake Corrib disappeared in the distance east as far as the eye could see. Mangus had drawn a bath for Marcus who really appreciated the warm, relaxing bathing.

Mangus said, "Mr. Daly, I can unpack for you and place the clothes in the wardrobe and linen press. You can just relax and make yourself ready for dinner. I'll take care of your clothes and room."

"Thank you, Mangus. I'll just be glad to let you take care of me, for all of a sudden I find I'm tired. I'll dress and meet the Guinnesses in the living room at 8:30," said Marcus.

At 8:40 Marcus entered the living room and Mr. and Mrs. Guinness rose and greeted him. The other guests at dinner were Mr. and Mrs. Donlin. They had a horse farm north and west of Cong and they would be hosts to Marcus in two days. They came to meet Marcus and determine what type of horse Marcus would be interest in. The group went immediately into dinner.

The dining room was furnished with Chippendale dining chairs around a mahogany dining table that seated fourteen, and a buffet, server and china cabinet of mahogany in the Chinese Chippendale manner including fretwork and scrolls. The dining room was serviced by two servants in black uniforms. The Chippendale china cupboard was filled with blue and white china that had belonged to Napoleon and was purchased in France from an antique dealer. They were cobalt blue, with portraits of the Bonaparte family on the plates. With the gold encrusted on the edges of the plates and serving pieces, the china cabinet glowed from the interior. The buffet held the silver service with coffee, tea, hot water, sugar and creamer. At each end of the buffet were candelabras with six branches holding candles lighted for dinner. The table had an Irish linen cloth with cut work and lace inserts and matching napkins. The centerpiece was a large blue and white tureen holding a mass of light and dark pink rhododendrons. The table had the Georgian candelabras with pink candles, and the china was white with gold trim with a dark pink band. The service plates were white encrusted with gold in a lace design that was dotted with pin sized rounds of gold on the edge. It was quite a colorful and cheery table for this night, and as the conversation commenced and the fire glowed, the room became warm and the conversation friendly. Mr. Guiness asked Marcus about his trip by sea to Ireland.

Marcus replied, "It was a most enjoyable voyage, and I met some interesting people aboard ship, so the time went quickly with the good company."

Mrs. Donlin asked, "Mr. Daly, is your wife going to join you here in Ireland? We would like to meet her, if so."

"No, Mrs. Daly and the children are leaving for Paris for the school year. She will come to Ireland at a later date, but this trip she had to give up. We have four children and the older girls are finishing their schooling in Paris while the younger two are being tutored at the Paris apartment," answered Marcus.

Mr. Donlin said, "Well, when you and Mrs. Daly do come again you will come to visit us. Since we will be going to our place with you, you will know the way, so no excuses for not visiting."

Marcus laughed and said, "That's right, and we wouldn't want to miss a visit with you, anyhow!"

The men's conversation then turned to horses, fishing and business while Mrs. Donlin and Mrs. Guinness exchanged compliments regarding each other's dress and the fine service of china, linens and silver. After the seven-course dinner they went to the living room to have Irish coffee. This was done in the grand manner, with the servant pouring the coffee from the silver coffee pot into the glasses with a flourish from a height that would make most people miss the glass. Other than the crystal stemware and the servant's individual ceremony, the coffee was made the same as the pub, with two sugar cubes, one jigger of Old Bushmill Irish whiskey, strong black hot coffee and double cream poured over the back of the spoon. The Irish coffee was served in a goblet the size of a wine glass.

In the living room the men had cigars, and the women made their way to the front sitting room where they could visit and not be in the room with the odoriferous cigars. This suited everyone just fine. They would join each other shortly before the evening ended, as there was to be typical Irish entertainment which consisted of Irish melodies played on the harp by a young lady musician. Since this day had been strenuous for everyone — Marcus' arrival in Ireland, the Donlins traveling from northeast of Cong, and the Guinnesses only coming from Dublin a few days before — the evening ended shortly after the entertainment and they all retired to their suites. Just before the evening ended, the men decided to fish on Lough Corrib in the morning. There were some mighty salmon trout in the lake and great sport for the fisherman.

Marcus made his way to his room and found the drapes still open. "What a sight," he thought. The moon was full and its brightness was reflected in Lough Corrib. The misty rain had left and the night was still. Marcus stood for a moment to look at the beauty and satisfy his soul. Sleep would satisfy his body aches, and by morning his body and soul would be restored.

Marcus awoke at 6 a.m. and took his toilet. At seven o'clock there was a light rap at his door and, without an answer from him, the maid entered carrying a tray with hot tea.

"Good morning, sir. 'Tis a bright day you'll be having today. The wind has stopped and no rain is in sight." All this was said by the maid to the room in general, for she never looked at Marcus or acknowledged that he was in the room. Generally the guests were still in bed and the maid's arrival was the wake-up call. The morning greeting to all was the same and she chattered on as she sat down the tray and flung the drapes open. Regardless of when one went to bed, the maid arrived at 7 a.m. As she turned from the window she saw Marcus standing, watching her. "Oh, sir, I did not see you. Breakfast will be served at eight in the morning room." She said all of this without eye contact except for the brief moment when she turned. Marcus was the first call, and the maid hurried on her way to deliver the tea to the Donlins. Master and Mistress Guinness would be served by the upstairs maid.

Marcus sat down to have his morning wake-up tea and took from his case some writing paper and penned what horses he specifically wanted to see at the Donlin's farm. He wanted some hunters and jumpers, and they had the finest. The thoroughbred in Ireland was a combination between the Arabian studs and Irish horses. The Irish horse had the benefit of moist mild climate with lush grasslands and the bone-forming limestone that the grass grew on, and it gave their bones great strength. Also, their life was one of the outdoor all year long, with climate neither too cold or too warm. Many armies had been fielded with Irish horses. The fame of Irish hunters as the best chargers in the world of racing and international show jumping was well founded and known to horsemen everywhere.

Marcus knew he would be proud to own the Irish horses for riding, racing, hunting or show. It was nearing 8 a.m. By the time Marcus found his way down the stairway and followed the noise of the voices in conversation, he was the last to arrive. They greeted Marcus and asked if he was ready for the fishing on Lough Corrib. They didn't need to go far as the Lough was at the door. The ghillies were ready and the three boats were close by the castle. Marcus, Mr. Guinness and Mr. Donlin were at the shore as soon as breakfast was over. It had been a buffet breakfast with eggs in a twelve-egg-cup silver server, bacon and ham and beef strips for meat, stewed tomatoes, sautéed potatoes and meat pie. No one went hungry. From the kitchen came a lunch for the fishermen that consisted of cold chicken, bread, cheese, Guiness stout, apples and cakes. The ghillies took the men to the boat. The ghillies had prepared the tackle boxes with the bait certain to catch the salmon trout of Lough Corrib.

Marcus couldn't believe how excited he was, and then he remembered he had not had time to fish in years. Actually, he didn't mind that he hadn't but thought it wonderful to have the chance to try to catch a lake trout. With a ghilli to guide him, he thought to himself, he might be more successful than in Montana, where the last fishing trip proved to be barren.

The three men each had a ghilli and their own boat, for they would fish different areas. Marcus hadn't discussed his fishing abilities with Joseph Donlin or Edward Guinness, and he certainly did not want to tell his ghilli what a poor fisherman he was. When Joseph suggested a contest on biggest fish, Marcus agreed to put up a wager, but he had no confidence that he would be the winner. The ghillies took their fishermen to the boats, seated them and started rowing to their respective areas. Marcus' ghilli rowed him to the south side of the lake, not too far from the castle. There were some small islands blocking his view of the others. Marcus took the rod from the ghilli and the ghilli rowed slowly while Marcus trolled. Then they stopped and Marcus wound the reel slowly and there was a sudden jerk on the line. "Easy, Mr. Daly, let the fish set the hook. By glory, you have a strike."

Marcus said, "Hey, that was some strike! I'm hanging on! I think I have a whopper!"

"Take it slow and easy and let him fight, but don't loosen the line," said the ghilli.

The ghilli started rowing with Marcus trying to reel the trout in. "He is a fighter for sure," Marcus said.

"You're bringing him in nicely, Mr. Daly. Hang on, I'm going to stop rowing and you bring him in. He's played himself out now. I'll net him, and when you get the fish close I'll club him."

The sweat was pouring off of Marcus. "If I lose him," he thought, "the ghilli might push me overboard." The ghilli was so excited and intent that the disappointment would be a terrible shame on Marcus. The ghilli had been rather cold in his acceptance of Marcus as his charge. He knew the other two were excellent fishermen and the ghilli was doubtful if Marcus Daly could compete. Marcus was aware of this and he hoped he wouldn't fail at the last minute. "By gad, here he is and what a big salmon trout. Mr. Daly, that will feed all the guests the first course and the servants, too. I've got him, Mr. Daly." With that, he swooped up the fourteen pound salmon in the net and clubbed it over the head for the kill. One might have thought Marcus was the greatest fisherman in the world. The ghilli broke out a bottle of Irish whiskey and offered Mr. Daly a drink. Marcus knew that he was a success. The ghilli was proud of him. And that is the way it is with a man and his guide in Ireland. They fished until the afternoon sun was high in the sky and caught a few more two- and three-pound trout. They set off toward the island to have their lunch. The other two boats were already there, but had just arrived.

The ghillies took the lunch baskets from their boats, spread the tablecloths and placed the seat pillows for Joseph, Marcus and Edward. They took from the baskets a three-piece setting of china and silverware and put them on the tablecloth for the three men. They opened the food basket and placed the chicken, cheese and bread in the center of the cloth for the men to serve themselves. After this ceremony the ghillies left the men alone and went off a short distance to prepare and eat their own lunch. No one ever knew whether the ghillies wanted to be alone or if it was the master's idea, but it was the custom.

After lunch and the discussion about the success of the fishing expedition, the time had come to see who won the fishing contest. Mr. Edward, as host, called upon Mr. Marcus as the honored guest to be the first to produce his catch. Marcus' ghilli came forward with his fourteen-pound salmon trout and three four-pounders, too.

Mr. Edward said, "Marcus, that old rascal has been around here for years and we've tried to get him. What bait did you use?"

Marcus replied, "My ghilli, Mark, baited the hook."

"Mark, what bait did you use?"

"It's one I made myself and it has lots of red and yellow and glitter. That old boy hadn't seen that bait before!" They all laughed and went back to the weigh in. Mr. Joseph had a ten-pounder for his largest and Mr. Edward had a nine-pounder, so Marcus was the winner and collected his money, which he in turn shared with his ghilli. It had been a great day for all, with no rain, great fishing and good fellowship.

They returned to the castle at 3 p.m. with just enough time to clean up and ready themselves for tea at 4:00 and cocktails at 6:00.

The next day would be spent at Joseph Donlin's horse farm outside the town of Cong. Mr. Guinness would accompany Marcus, for it was always a pleasure to see horses of such stature as those on the Donlin horse farm. The horses were bred for the hunt and steeplechase, as well as for carriage and riding.

The day was misty and the sun could not break through the overcast sky, but Marcus and Edward Guinness were not deterred by the weather and were ready to leave soon after breakfast. The coachman brought the covered coach to the front of Ashford Castle. After the breakfast of steak and eggs, hot breads and fruit, they would not need to stop until they reached the Donlin's in two hours. Marcus relished the misty rain, for he knew it made Ireland green. It was a necessity for the farmers' crops and for life itself. Marcus mused that civilization, with all its advances in the arts and sciences, owed its existence to a six-inch layer of topsoil and the rain.

"Marcus," said Edward, "what are you looking for at the Donlins? I understood you wanted race horses. Their horses are for jumping, hunting and show riding."

"I know, Edward, but also I want the finest riding horses, for we have a vast area to ride and hunt. I'm making my own Irish country living at Riverside in Montana. My girls and son are riding now and I want them schooled in dressage, carriage driving and riding to the hounds. That is one part of their education that I will undertake myself. Of course Mrs. Daly, herself, is an excellent horsewoman.

"Thoroughbreds are excellent riding horses and performers, but they are very temperamental. Warm bloods like Mr. Donlin's are calm by nature. One can train them while showing them. Warmbloods have a natural rhythm for dressage and are able to keep their minds on the different movements. They are dependable for carriage horses, riding, jumping and hunting. I understand Mr. Donlin has the finest, and I'm hoping to obtain five from his farm. I will go to Tully in County Kildare to find the thoroughbred race horses for my farm and then on to England and France. Actually, I'm going to purchase a breeding farm in England so that the mares can be bred by the best stallions in Europe. That way I can build my stable with Europe's finest and America's finest at the Bitterroot. I don't intend to race my horses in Europe but in the U.S. I'm having race tracks built at Butte and Anaconda. My partner, Mr. Haggin, has some fine horses in California and races his stable there and will bring some of them to race in Montana.

"You asked me a question, Edward, and I got carried away. You probably heard more than you wanted to know, but I am fired up about this project. I want it to be my finest endeavor. I never thought it would be possible that this Irishman could have within his reach a great racing stable."

Edward reassured Marcus that he was glad that Marcus felt at ease to confide in him. "Marcus, you are an exceptional person. You have accumulated great wealth in America and I believe you will be as famous for your racing stable as for your copper mine!"

The carriage turned into a lane and followed the hedge row to an open area with the manor house at the center of the grounds. Behind and to the side were rock walls and hedgerows marking off different pastures. Into view came the horses, running and snorting, with mares and colts in one field, stallions in another. Marcus felt his heart pound in his throat. What a sight!

The carriage pulled in front of the manor house and Mr. Donlin came to the front steps while the butler took care of the step-down from the carriage. "Marcus, Edward, welcome to Mayo farms. The horses are out and running. I want you to see them in the fields before we show them under tack."

"We saw them as we drove into the manor," said Marcus. "It was a thrill to see such magnificent animals."

"All right, gentlemen, then come in and we will have our lunch. After that, we're off to the show ring in the barn." The men were excused to get ready for luncheon.

The lunch was served from a buffet laden with whole chickens, leg o'lamb sliced thin with mint sauce, baked caramelized potatoes, onions and carrots. Dessert consisted of sponge cake soaked in rum, with whipped cream.

The show ring was ready by 1:30 p.m. and the men entered the arena as the clock struck the half hour.

Joseph Donlin sat between Marcus and Edward. "We will show the hunters first. We bred our great Irish hunters, carriage and riding horses called cold bloods with Arabian stallions hot bloods. The hot blood of the Arabian and the cold blood of our strong, steady Irish horse has produced a warm blood that has the best of both breeds. They are steadier than the thoroughbred and have more endurance. They can carry more weight for a longer time.

"The hunts in Ireland are arduous, and our hunters can carry you to the hunt, run and jump up the high banks of the ravines, jump the hedges and, at the end of the hunt, they will still be able to carry their rider back to his home.

"Here are two of our finest. Marcus, take a look and tell me if I exaggerate," said Joseph.

Two fine bay horses came into the ring. First they were walked around, then trotted, then cantered, and then the jumps started. It was flawless, with the riders and horses as one. Without hesitation, the horses, one after the other, jumped higher and broader obstacles. When the horses had finished their jumps, Marcus and Edward stood up and applauded. Joseph knew his horses were exceptional and they had performed in masterly fashion. There were more hunters shown, but Marcus already knew he wanted the first two he'd seen.

Next came the steeplechase thoroughbreds. These horses were like race horses, for they ran timed races with obstacles and over rough terrain.

After the last horse had been shown, Marcus said, "Joseph,

you have a fine horse farm here. I would be proud if you would allow me to purchase the first two bay hunters shown and the two steeplechase thoroughbreds, #4 and #6. I don't know what your asking prices are, sir, but I know these are the finest and I'm obliged to pay what you ask."

"I can't think of anyone I would rather sell my horses to nor could they be better treated than with you as owner, Marcus. I will need to look at some books on each of these horses and I will get back to you in an hour with their prices. In the meantime we can go to the tack room and have a hot toddy.

"Sorry, Edward, but stout is not my favorite. I prefer Irish whiskey."

"Not to worry, sir. Enough people drink Guinness Stout that we do not worry about your preferences." They all laughed.

The tack room was attached to the thoroughbred barn and held the ribbons and trophies of Joseph Donlin's horses. As one read and looked, the story and lineage of the horses evolved. One could imagine the years of work and selective breeding it took to produce these fine horses.

Edward and Marcus sat at one of the tables after examining the trophies, pictures, paintings and stories. Joseph went into another room to look into his files and the records of the horses. When Joseph returned he presented Marcus with the prices of the horses selected. The price was high, but to Marcus it was a bargain, for he had never seen such horseflesh and he wanted them. So the deed was done. The money was paid. The horses would stay at Donlin's until Marcus could get his horse farm in England. This was where he planned to gather the Irish, English and European horses to breed the finest, improving and selecting and then sending them to Riverside. Here Marcus would have the results of these great horses he had chosen overseas.

Edward and Marcus spent the night at Donlin's, for the roads were treacherous at night and there was no reason to endanger horses or men to get back to Ashford Castle. The Donlin's hospitality was cordial and warm. The country manor house had large fireplaces in each room burning with peat that added a particular glow and warmth to the room. Marcus was not hungry at dinner. The excitement along with several drinks and much good food all day had filled his being. After dinner and a short visit with his host, Marcus retired to his room thinking of the next stop on his quest for the finest horses. He would be at the Guinness' for few more days, then he would take the train from Galway to Tully, County Kildare.

Marcus went down in the morning for breakfast and entered the dining room as Joseph came in. "Marcus, are you all packed and ready to go? I would like to have you stay longer. We have good fishing and hunting here, and the neighbors would like to meet you."

"Joseph, it has been grand meeting you and being with you, but I'm a man always in a hurry. I want to get my stable of thoroughbreds purchased, trained, and able to race within the next few years. I must be on my way to cover the European market. Tell your friends I'm honored that they care to meet me. I'll be back; and, Joseph, it would please me if you came to visit in New York and Montana. It would be a real pleasure for me to have you as a guest."

"Well, Marcus, I don't think I will get to America. Actually, I can't go anywhere. I'm the sole person I trust to manage my horse farm." Joseph had to laugh at himself, for what he said was true.

Marcus said, "Joseph, I know how that is. I found it very difficult to allow anyone to take responsibility from me. But, as more business came under my care, I had to find good men and trust them to do the task. When you get yourself a good worker and manager, take some time for yourself and your wife, and come to see me and my family at Riverside in Montana. I'll have a fine horse farm there shortly and you can tell me if I did it right!"

After breakfast the coach was readied and Edward and Marcus said their goodbyes and thanks to Joseph Donlin. Mrs. Donlin had not come down yet but had said her farewell the previous night after dinner.

On the journey back to Ashford Castle, Edward Guinness and Marcus Daly reminisced about their past lives and then settled into a quiet time on the last part of their journey. As they neared Ashford Castle, the horses seemed to pick up the pace. The warm barn and a ration of oats would be theirs at the Ashford stables.

Marcus said, "Edward, the pace has quickened. The horses are headed for the barn. We'll be at Ashford in a minute."

The horses entered the tree-lined lane by the river Cong that flowed out of Lough Corrib and into the Atlantic. "'Tis a glorious sight," Marcus said. Edward certainly agreed, for Ashford was the Guinness family home and had been so for generations. The castle and surroundings were part of Edward's life, and it was here he felt most content and happy. The next day he would walk with Marcus and show him the Abbey ruins on the estate. They entered the front door and parted until dinner. During dinner Edward asked Marcus if he would like to walk to the old brick arched bridge that led them to the Abbey ruins. Marcus was glad to have a plan for the next day that included a little exercise. He had ridden horses here, but a walk and a chance to see the area on foot would be good.

The next morning after breakfast, Edward and Marcus dressed in plus fours, wool socks, brogans, Irish tweed jackets and, carrying walking sticks, they started off from the castle toward the river Cong. A mile of walking brought them to the bridge. It was wide enough for a horse and rider, for a person walking or a horse with a small cart. Across the bridge loomed the ruins of the abbey. Large exterior walls that had nothing inside except some crumbling stones laying amidst the grass and vines that had claimed the floors and part of the walls. These abbeys were all over Ireland and the learned monks from the golden age of Ireland had provided knowledge and teachers for Europe and England as well as Ireland. Marcus was not much interested in the ruins, but he wanted to know the history of his birthplace Ireland, and the ruins of the abbeys, castles and cities told a tale of their own, depicting the history here at Ashford and in the other counties Marcus would visit. Anxious to get on his way to Tully and the center for thoroughbred racing horses, he asked Edward if he might leave Ashford Castle before breakfast the next morning. "Edward," Marcus said, "could your man drive me to Galway and start earlier than planned? I will not have much time between arrival in Galway and boarding the train to Tully."

"Of course, Marcus, not a problem at all. We will have the kitchen make a lunch and hot drinks for your trip. That will carry you over until you reach Fox Inn."

The next morning the coach pulled to the front, ready for

Marcus to board and be away before 7 a.m. Marcus rode alone in the coach while the driver and out rider were up on the front of the carriage. Because he had no one to talk to, Marcus noted the countryside and particularly the sheep "woolies" so white against the green grass and blue sky. The clouds flitted by, leaving spaces of blue sky dominating the horizon.

The left over rain drops sparkled on the greenest of all grasses and it looked as if emeralds were scattered throughout the land. Marcus wondered why he had not remembered the beauty of Ireland? Or was it because his life here had been such a hungry time and hard for his family or was it because youth takes for granted scenery they have always seen? Bidding farewell to Edward had been a pressing of hands and a "Please return and come to the Bitterroot, or at least New York," but Marcus was surprised that he felt a loss as he bade Edward farewell. After all, they had really not known each other before. Now Edward and Marcus were kindred souls and a strong friendship had resulted.

The carriage made a quick stop at the Fox Inn. Marcus said to the driver, "Michael, we have enough food aboard let us just stop for a refreshment."

"Yes, sir, Mr. Daly. We won't need to linger long and I know you are anxious to get to Galway."

The horses snorted and strained at their harnesses while the men were boarding the coach. They, too, seemed anxious to get to Galway.

The coach pulled up to the railroad station where a porter picked up the luggage. Marcus started for the ticket counter with the porter and the luggage close behind. The thrill of the chase sent a charge through Marcus' being. He was going to Tully to see the finest thoroughbreds in the world. The Irish National Stud at Tully encompassed 100 acres, and its equine tenants were among the most valuable on earth. The horses' stalls had brass plates with their names, number of races won and the amount of money won in their careers. Marcus wanted some of these marvelous thoroughbreds for his Riverside stable. The breeding was strictly controlled by the Irish Horse Breeders Association and, like other countries, they guarded their breeding secrets well. Each country believed it had the best horses, and there was great rivalry. Tully was the next station, and Marcus moved closer to the window. The train whistled and sounded a warning as it powered into the Tully Station.

When Marcus stepped down from the train he spotted a signboard held aloft by a messenger. The sign had "Mr. Marcus Daly" printed on it. Marcus hailed the carrier and identified himself, and the young man led him to a waiting coach.

"This way, Mr. Daly. Your agent, Mr. Lancaster, from London, is waiting at the stud farm for you. I will accompany you there and then take your luggage on to the Tully Inn where arrangements have been made for your stay."

"Thank you, young man. I am anxious to be on my way." The horses trotted off at the urging of the driver, and they arrived soon at the stud farm.

Mr. Lancaster was the agent Marcus had contracted to purchase a breeding farm for him in England. The final purchase would be done when Marcus arrived in London. The bookwork, the examination of the deed and the survey of the property had been completed. These preliminary works were a task, for the lawyers were precise and slow-moving in the closing of properties.

To give them just due, England's land records went back centuries through generations and it did take time to trace and find the facts from so many years and numerous owners. Marcus was glad he hadn't been involved with all that, for patience was not a strong part of his character, and the English wordiness, slowness, and attention to petty details would have driven him to distraction. However, he did have to smile at the last letter he had received from the English law firm of Wilson, Bailey and Endecott. The letter informed him the firm would now be known as just Baily and Endecott since Mr. Wilson had passed away in 1852. Since this was the year 1890, Marcus thought it a typical example of the slow and ponderous movement of English firms.

"Ah, here we are at the breeders' sale of the thoroughbred racing horses, and Mr. Lancaster will be in the #6 box at the show ring," said the messenger. Walking briskly, Marcus entered the ring. The smells of horses sweating and the dust rising in spats as the horses were guided through their paces caught Marcus' subconscious thoughts and brought to him fond memories of sale barns and the race track paddocks. He paused for a moment to watch and then briskly went up the stairs to box #6. "Mr. Lancaster," said Marcus, "I'm Marcus Daly. It is good to meet you in person instead of knowing you only by your pen and paper."

Mr. Lancaster and Marcus shook hands and then Marcus said, "Let us get to our business. Have you found any racers that we need to buy? If you have, I'll take a quick look and leave the final purchase to you. I'm anxious to get on to London and finalize the purchase of the breeding farm."

Just then two of the horses that Mr. Lancaster wanted Marcus to see came into the ring. The horses moved wonderfully and their coats glistened. These horses at Tully lived better than most people and they were pampered to a fine point. The horses were strong and were expected to race on dirt barely leveled and at full speed for distances of two-and-a-half miles, much further than the one- to one-and-a-half mile races in America. This ability to run at full speed for longer distances made the Irish horse the strongest of the thoroughbreds, with the greatest endurance.

Marcus liked what he saw and asked Mr. Lancaster how many horses he needed for breeding. "I want five of these Irish horses to breed with other stock from England and Europe"

"That's fine. Buy them and let us be on our way to our rooms at the inn."

Tully was a typical Irish town with white washed stucco buildings, winding streets and wooden signs hanging over the walkways. The main artery was crowded with horse-drawn carriages, pedestrians and dogs vying for the small space between street and building.

Marcus and Mr. Lancaster agreed to meet for dinner at seven. They would dine at the Tully Inn, for the cuisine reputedly was excellent and there was no need to search for better fare.

The pub's entertainment was provided by its patrons, and when one was performing there was a polite silence, and close attention was paid by all as they listened to the song or poem. Most of the time it was whole groups of jovial drinkers joined in song and laughter. As Marcus seated himself at the bar, the owner of the pub asked Marcus his drink order. Marcus ordered a Guinness Stout and engaged the pub keeper in conversation. "Sir, you have a very interesting pub and it is furnished with fine details. Is there a story about your pub?"

"Yes, sir, this is the oldest and finest pub in Tully and I'm proud to be the owner. 'Tis the most expensive, but it has been worth it. Look at the fine carpet, patterned with flowers and vines -- not a bare floor with sawdust. And the carvings on the mahogany bar are specially done by a fine wood carver. The Celtic symbols on the bar posts to the ceiling are worth looking at, sir. This pressed tin ceiling has been here seventy-five years and still is in good condition. Mind you, I take good care of it by washing it down and using a little linseed oil in the water," said the pub owner.

Marcus looked at the room and again at the bar. The bar was mahogany with three arches across the top setting off three mirrors. The mirrors were partially obscured by the various bottles the bartender needed. It was decorated more lavishly than most pubs, and Marcus was delighted that he had stopped for refreshment there. Marcus was a man's man and he was completely comfortable with the boisterous patrons. He introduced himself to the young men beside him. "Are you familiar with the National Horse Breeding Operation here in Tully?" Marcus asked.

"Of course, sir, anyone who lives in Tully knows about the breeding farm. It's the finest in the world as, of course, is the Irish horse."

"You are right on that last statement, for I've just come from viewing the finest horses I've seen — ever. I'm going to take some with me to the United States and they will be the start of my racing stable. God willing, I will show all of America what great horses the Irish have." That was the right thing to say, for everyone gave a cheer and toasted Marcus and his quest for the Irish horse and his plan to have the finest stable in the world. The hour passed quickly. Marcus said his farewells to his new-found friends and shook hands all around. As he left he gave a generous sum to the barkeep to buy drinks for the men.

Marcus entered the Tully Inn and went to the desk clerk for his key. There was a note in his box which the clerk handed to him. His heart pounded, for it was a letter from Margaret. Even though Marcus had been busy and had enjoyed the company he had shared, always in the corner of his mind was Margaret and the children.

"Marcus, my dear,

We have finally arrived in Paris. I thought the sea voyage would never end. The seas were rolling and not too rough, but it is an eleven-day voyage. The children, Harriot and Marcus, had a grand time. The steward organized games and toys for play, and each day Marcus and Harriot could bring a treasure to show and tell their new friends and vice versa. Madge and Molly found some young people (I approved of them as I asked who they were from the purser) and amused themselves. But, I — I was lonely Marcus, and longed to hear your voice and know that you care -- I do hope it is not too long before you visit Paris.

We are quite well settled. The girls go to the Elysee tomorrow for their French lessons. It isn't far from our apartment.

The little ones are being tutored and they are happy to go to the park after studies. Paris is such a lovely city and although it is cold here, it is not as severe as New York.

My dear Marcus, take care of yourself, eat properly, don't drink too much Irish whiskey. Get your rest and take time to enjoy yourself.

I have your itinerary, and there will be a letter for you in London.

I know you are going to Cavan to see Honoria and to visit your old home. I'm so glad you came to America. I'm so glad I went to the mine with Father to meet the mine manager. I'm even glad I stumbled and fell in your arms. (I knew then that I would love you forever, Marcus).

Until Paris, dear one, I will see you in my dreams. Your faithful wife with,

All my love,

Margaret"

Marcus immediately sat down at the writing desk and wrote a letter to Margaret.

"Dear Margaret,

A boatload of love hit me when I read your letter. You are dearer to me than you can know, and it seems unfair that our lives have different directions.

There is so much to be done, my dear, and I am trying to accomplish my tasks and desires so that we will have our special time together soon. Then we will be able to be together to enjoy our life which will have less work and more play for both of us. Make one of your famous lists of 'wants' for you and me.

I am excited about the horses in Ireland for I think I have seen the best horse flesh ever. Tully and the National Stud Farm were unbelievable. It was an education as well as a pleasure. I have bought some excellent horses, including a fine horse for you, Margaret. You will enjoy riding and owning her.

Tomorrow, I will go to Cavan and visit Honoria. Then I will take the train to Dublin and from there the boat train to London.

Mr. Lancaster came from London to Tully and I will meet

him again in London to finish the purchase of the horse farm in England. The farm is south of London, which is convenient to Southhampton where the horses will board the ships to America and to the Bitterroot.

This horse business seems to grow bigger and bigger, but I can't help it, Margaret. This enterprise is my finest endeavor, and it is such a pleasure. I'm a true Irishman, and fine horses are our passion.

There is one plan I would like to make. I will make arrangements when I am in London for the Ascot Races the second week in June. That means we will be arriving late at Riverside, but I do want you to join me for this occasion. Plan on it Margaret. I'm sure you will need to see your dressmaker so that you can be at your best for the Ascot.

Tell the children "Papa loves them" and to be good and I will be there soon.

My dear one, I miss your sweet smile and the comfort of your arms.

Yours faithfully with
My love,
Marcus

After writing the letter and posting it, Marcus had dinner and retired early. He would take a coach the next morning to the train that would take him to Cavan.

Chapter Seven

The coach left at 7 a.m. and Marcus was there waiting when the carriage horn blew announcing the approach of the coach. It was a touring rig pulled by four high-stepping horses. The driver was up front with a second man behind who helped load the luggage and the passengers and would blow the horn on arrival and departure after the coach arrived at the train station. The ride was still two hours and the early departure left the passengers either silent and thoughtful or nodding off for a few more minutes of sleep. The train to Cavan was not very full, and Marcus was able to choose a window seat where he could see the countryside. The closer he got to Cavan the smaller the farms became. The animals were mostly pigs and sheep. The houses were white-washed stucco with thatched roofs and small neat yards. In the summer he remembered his mother planting a few flowers in the boxes at the front window. Although the Irish were oppressed and the potato famines had driven many Irish from Ireland, Marcus remembered that, even so, they somehow could find happy times with fellow Irish singing or telling tales which brought out roars of laughter. The Irish wit was a true trait of the Irish, and their humor came from the unexpected punch line. Perhaps this ability saved their sanity and made life bearable. Marcus remembered all of this, and, of course, Butte and Anaconda had large Irish contingents and they hadn't lost their humor by traveling to America.

Cavan station came into sight and the conductor announced the arrival of the train into Cavan at 2 p.m. Marcus had chosen to eat a light lunch on the train, so he went immediately to a hack and engaged the driver to take him to his former home outside Cavan. Whether Honoria was expecting him this day, Marcus couldn't know for he had told her his arrival would be that day or the next. He instructed the driver, Turn here and go to the 6 kilometer post and turn right."

As they came up to the house with its tidy yard and lace curtained windows, the door opened and there was Honoria, Marcus' oldest sister.

"Marcus, you're here! I would not have known ye since I haven't seen ye since ye were fifteen. Haven't you turned into a fine man. The Daly Clan basks in your glory! Your success has came back to the ears of all in the area. Come in, I have the kettle on and some Irish cakes so that we can talk and remember when."

"Honoria, I would have known you. Your hair is grey, but the face is the same. I'm glad to know you are in such a fine fettle. Your brothers and sisters in America are anxious to hear news of you."

Marcus paid the driver when his luggage was unloaded and told him to return the next day to pick him up. "Here's a sovereign, lad to make sure you remember. I want to take the 3:00 train to Dublin so be here by 1:00. I will count on you."

The driver said, "Thank ye, sir. I will be here by 1:00. You can count on me."

Honoria and Marcus, each carrying the luggage went into the house where Marcus, Honoria and the others had been born. Marcus thought of his mother as he entered the door. It was she who had given him the only education he had had, teaching him to read and write and inspiring in him the thirst for knowledge. He wished that she could be there, to be proud, to give him a hug. He thought of his father who had worked hard in the fields to provide for his family, who had encouraged him to go to America to pursue a better life, and who had given him the gold piece that he never spent but had carried always as his talisman. Marcus was glad he had been able to help this man who had taught Marcus to value honesty and hard work.

Marcus could have dwelt longer on these somber thoughts but Honoria would have none of it. She was a small wiry woman with bright blue eyes and dark brows like a bird's wings. Though her hair was grey, her eyes were young and expressive, twinkling with laughter. When she was sad, they became dark forest pools full of Celtic emotions. Honoria had the tea and cakes on the table in a flash, and she and Marcus talked of their life before America and after. Marcus felt very much at home, for blood is thick and it is easy to be with one's kin.

She was clearing the table when a knock came at the door. Some of the neighbors had come to see Marcus, for they were curious about the sort of man he had become. The group came into the living area, spilling over from the living room into the adjoining kitchen and dining areas.

"Marcus," said Donovan, "do you remember when you use to take care of my pigs? I tell that to all for I don't want you to forget and become haughty. You are just Marcus, my good hired hand to me."

Marcus laughed, "I remember you well and I remember those pigs. Actually I have never tended pigs again. We do have some at Riverside, my home in the Bitterroot Valley of Montana, but I'm not inclined to visit them. I'm a horseman and prefer their company to any other in the animal kingdom." And so the afternoon went with everyone remembering old times and telling or listening to the tales of all that had happened in the intervening years. Marcus enjoyed the banter and he appreciated what good, honest people they were. He wished he could do more for them. He thought of sending some pigs from England. They had been bringing out some new breeds, and Donovan and his neighbors might improve their herds with new stock. He decided to do that.

The evening meal of simple Irish food was as satisfying to Marcus as any dinner served at the Ritz. Honoria served warm Irish soda bread with freshly churned butter dripping from it, and pork roast cooked on an open rack in the fireplace, with potatoes and cabbage. It was a treat for Marcus. During the evening Marcus asked Honoria why she didn't join the rest of the family in Butte and Anaconda or to live with a favorite sister in California. Honoria said, "Marcus, I'm Irish. This is my house, I have no needs or desires that aren't fulfilled here. You have been generous to me, but your monthly stipend to me is not really necessary. I take it because you insist and it makes my life very comfortable. I love Ireland, unlike you who had to leave, you have made it possible for me to stay. Thank you, dear brother. I say a prayer for you and yours everyday. A better man never was, Marcus."

What could Marcus say to that? Marcus mumbled something about his duty and his affection for her and the subject was dropped. He continued to answer Honoria's questions about their brothers and sisters in America until the hour was very late.

Next morning Marcus was up early, but Honoria had risen even earlier, and the eggs and bacon were popping in the skillet. After breakfast he walked around the farm and was pleased to see that everything was well tended. The hired man was taking care of the chickens, pigs and cows. Marcus could understand why Honoria wanted to stay on the family land. It was peaceful here, and she had hired help, which relieved her of hard labor. Honoria had all she wanted and was content. "Isn't that all anyone strives for?" thought Marcus.

Encountering the hired man, Marcus said, "Miss Honoria said your name is Mike Timlin and that you take care of the land and stock with great pride. I thank you for it. I'm Honoria's brother, and I'm on a short visit."

"Yes sir, Mr. Daly, I knew you were coming and I'm glad I got to meet you. You have been good in sending suggestions for rotating the crops, and I do see a difference in production. Miss Honoria is kind and generous to me and allows me to raise a few animals along with hers. Do you know, sir, that Honoria has 'The Sight'? People come from afar to have Honoria heal them. She knows things no one else does. She is needed here for our welfare. I know that is why she never left for America like you and the rest," said Mike

"My heavens, Mike. She never has said anything about it to me. But it's good that she is needed," said Marcus. "I'm glad you have your own animals here at Honoria's, Mike. Just don't get so much you want to leave. We all depend on you. By the way, I'm going to send some new breeding stock for pigs and maybe some sheep from England. I will send you information about the breed I send."

"Very good, sir. I will be glad to know more about the animal stock. You need not worry, I won't leave Miss Honoria on her own. Me own wife is a good friend to her, and Honoria is a good friend to us."

Marcus took out the book he carried in which to write pertinent information for future reference. He jotted down a note to get livestock for Mike and for Honoria's friends and neighbors.

One o'clock came in a hurry, but Marcus felt his visit was complete even though it had been brief. Honoria wanted Marcus to stay longer, and a tear had even rolled down her cheek. Marcus said, "Now, Honoria, me dear, I will write to you. You have all your good friends and neighbors here and you certainly can come to visit us anytime so 'tis not goodbye."

"Marcus, I haven't seen you in thirty-four years, and then you are off after only one day. But I am grateful for the short time we could be together. May God be with you, dear brother. I will say a prayer for your safe journey."

The horse and carriage and driver turned into the lane promptly at one o'clock. Marcus was ready and the driver turned the horses around. As soon as the coach door was closed with Marcus seated, they were off to Cavan. Honoria waved to Marcus until he and the coach were out of sight.

From Cavan to Dublin Marcus took the same railroad route that he had taken as a fifteen-year-old. It had been a thrill and a frightening experience for a fifteen-year-old boy to leave his family forever. He would not linger in Dublin but would board the ship as soon as he arrived. The countryside looked the same with forty shades of green set in blocks, sagging clouds down on the hills and a misty rain. It was just as he had remembered. London was his next destination. The purchase of the horse farm near Epson would be signed on Thursday. This was Tuesday, and Marcus would have these three days in London on his own. He couldn't go out of the city, as he might need to see the lawyers about last-minute details. He didn't know anyone other than Mr. Lancaster, so Marcus decided he would look at the London sights. On Wednesday, the day would be spent touring London. He engaged a carriage to take him around the city.

"What a fabulous city London is!" thought Marcus as he took a walk after lunch from the Connaught Hotel. The Barclay Bank was in Berkley Square. Marcus couldn't understand why the American banks didn't give overdrafts as the British and Irish banks did. Marcus had an unlimited amount that could be used by him without that sum in the bank. He had the Butte bank do that for him, but the bank officers were not at all pleased. Marcus told them of the system in England, but the bankers were not impressed.

Marcus continued past Barclay's Bank on to Regents Street. The architecture on Regents Street was Georgian with classical details in the elements of the buildings, columns and pediments that had been inspired by the rediscovery of Pompeii in the 18th century. Marcus liked the look of the street's architecture. He walked by the shops but didn't feel confident enough in his good taste to buy Margaret or the children any presents. He criss-crossed the streets enjoying the sights and ending up on Picadilly. There was the store of Fortnum and Mason with foods and general goods served by men in formal morning clothes. Across the street from Fortnum and Masons was the Burlington Arcade. The Arcade had an ornamental iron and glass roof that covered half a block. The wealthy tenants next door could shop without braving the weather, for the Arcade abutted against the west building in the group and the patrons could enter the Arcade from that passage. The Arcade had several small shops: a greengrocer, a bakery and a butcher at one end and yard goods, jewelry, tailor shops, and fine leather goods at the other. The Arcade was not only functional, but also beautiful, with lots of light from the roof and chandeliers.

When Marcus came back to his hotel, the mail from the United States that had been held for him was delivered. He was glad he had taken his walk, for the late afternoon would be better for reading, as it would keep him occupied until dinner and leave no time for loneliness.

Most of the mail was from the Anaconda Copper mine in Butte and the smelter in Anaconda. Amazingly, all was going well. There had been a depression a few years previous, but the copper price was normal, and the hard work had paid off. Marcus took a break before reading the news from Riverside. All seven farms were established and he had personally chosen each foreman. He hired the best available men and paid them better than any other employer. He was very pleased with the foremen, and the glowing reports made him a happy man indeed. Since he was buying thoroughbreds now, he was interested in the report of the accommodations he'd planned for them. The veterinarian and an assistant were on the premises in housing built expressly for them. A veterinary clinic was being built, for there was no place to take ailing animals around Hamilton. Marcus' philosophy was, "If you need it, get it", and his horse operation would be unique. The horse farm would be ready for the new tenants from Ireland and Europe when the time came for shipping them. Again the nagging thought intruded, that if things were going too smoothly, disaster was bound to follow. This flashed through his mind, but he turned it back and decided to enjoy the present.

The following day, he would sign the papers to buy the horse farm near Epson. The group involved would meet in the afternoon after Marcus had seen the farm. He had seen pictures, but he wanted a walk through as well.

Since it was too dark for a coach ride around the Parliament building, Westminster Abbey, Tower bridge and the Tower, he would have to see them coming and going from London.

Epson was a small English town that turned into a exciting, busy, and entertaining city twice a year when the race meeting was held at the track at Epson Downs. The famous Derby was attended by crowds from far and wide. Marcus took the train to Epson. It was only a short distance from London and he was met by Mr. Lancaster at the station with a horse and carriage to take them to the horse farm. After viewing the farm they would meet their lawyers and the real estate persons in Epson.

"Mr. Daly, here I am," called Mr. Lancaster.

Marcus waved and approached Mr. Lancaster. As they shook hands Marcus said, "Did you leave Ireland all right? And did you leave the money there for my fine horseflesh?"

"Of course, Mr. Daly. They are outstanding horses, and I bid high enough to get them."

"Good, now let us see how you did getting a horse farm. The price was a good quote, and the location has an easy access to Southampton and the overseas shipping docks."

They turned together, boarded the carriage and quickly left the town of Epson behind them. Epson is in Sussex where the farms are manicured and lush. They turned on a lane close to the Downs race track and entered into an area confined by a three-rail white fence. The buildings came into view. There was a horse barn with ten stalls, a Tudor-style farm house and another small barn for other livestock.

Marcus said, "Mr. Lancaster, this looks as if it were made to order. I hope the house suits you and that your family will be happy here. If you need some redoing of the buildings, including your house, I want you to do it. My investment depends on you, and I want your life to be comfortable so that my profit will be!"

Mr. Lancaster smiled, "Thank you, Mr. Daly. It isn't often that one finds such a generous employer who gives such responsibility to an employee. I appreciate your trust in my judgment, and I assure you your investments will be taken care of as if they were my own."

"Let's go find those lawyers and the deed will be done," said Marcus.

As slow as the progress had been on the papers (it had been five months since their inception), the work was finished, and the signing and the shaking of hands was over in forty-five minutes. Mr. Lancaster was to be congratulated. There wasn't a last minute "glitch" which could easily have happened if all the papers hadn't been in order. Mr. Lancaster had done his homework.

Marcus said, "Mr. Lancaster, I would like monthly reports from you. Since it is close to February 1st, your first report can be March 1st. By that time the horses, and the repairs should be in place."

Mr. Lancaster said, "The horses will be, sir, but sometimes repairs take longer. I will stay right on the workers to finish quickly. I hope you have some luck in Europe. The Germans have some fine trotters, you know. Just send the horses here and I will take care of them until they can be shipped to you at Riverside."

Marcus boarded the train, not even taking time for lunch. He could eat in the dining car, and Mr. Lancaster had much to do and would get a "Bubble and Squeak" at the Epson Down Pub before it closed at 2:00 for the afternoon forced closing of three hours.

When the train arrived in London there was still enough day light for a sight seeing tour on the way to his hotel. Marcus caught the eye of a cabbie and told the driver to make a trip by the Tower, Tower Bridge, Trafalgar Square, St. Paul's, Westminster Abbey, House of Parliament and other grand tour sights.

The driver was delighted to do so for this would be a good fare and he could retire his hack and horse early and have an evening at home with his good wife. He not only drove Marcus, but also gave a running commentary on the sights and sounds of London. Marcus thought this was a great way to tour -- no walking, no people and no need to go into the buildings. When the children asked about London and places of interest, Marcus would be able to tell them all about the points of interest. He wouldn't tell them of his skipping the bulk of the lecture and tours.

The last place the driver took him before the hotel was London Bridge. Marcus knew the children would want to know about it because of the nursery rhyme. It was not as an impressive a bridge as Tower Bridge, which had the large towers that housed the mechanics for the drawbridge. London Bridge was just a bridge but at one time had stalls and shops and was ready to fall down -- thus the song and children's game. Having taken the scenic tour including the theater district and Picadilly Circus, Marcus felt as if he had seen enough of London. He would wait until another visit to visit the environs of London. Marcus was anxiously awaiting the next morning, because then he would take the boat to Holland, then to Germany, and then to Belgium on the quest of the finest horses in Europe. Then he would go to Paris, France, for a reunion with Margaret and the children. It had been a month since Marcus had seen Margaret. Although in the past he had spent a month away, he felt this time it had been longer. Maybe that was because he had seen and done so much and wanted to share these travels with Margaret in conversation. He wished Margaret had been able to be with him on this journey to find the horses.

He would be in Holland, Germany, and Belgium for three weeks and then he would go to France. The horse farms in these countries were numerous and each one had their credentials, ribbons, prizes and the lineage of the horses on display. The horses were shown by excellent riders. The owners of the farms had sumptuous feasts prepared for the buyers. The entire family of the horse owners and their friends also came for the festivities.

Marcus had agents in each country to introduce him to the horse owners and help him in selections. The agents had already made arrangements at the finest horse farms, and Marcus was able to buy both trotting, standard and racing thoroughbreds. In all, Marcus purchased ten thoroughbreds on this trip. His agent would watch for other prize winning horses throughout the years to come. Marcus had faith in these agents and never felt the need to do this personal buying tour again.

Marcus wanted some Belgian work horses for the farms at Riverside. The Belgian is the biggest and the most powerful of the working horse, and can weigh up to 2000 pounds. The horse farm in Belgium was owned by Donald and Joan Sontag. Marcus had written to them personally and had made his own arrangements with them. The time had come for Marcus to leave Germany and go to Belgium. On Marcus' arrival in Brussels he was met by Donald Sontag. Donald was tall, slim and in excellent physical condition. Marcus felt quite dumpy beside him, but then Marcus hadn't been handling 2000-pound horses, which was a real physical workout.

"Mr. Daly," Donald called, "over here. How are you? It is good to put a face with your name. My wife is anxious for your arrival and has spent the day fussing over, linen or cotton sheets, which china to use for dinner, what dress to wear or should she be in casual or more formal wear and so on. I'll be glad when you are in our home and she can no longer fret."

Marcus had to laugh, for he could tell Mr. Sontag adored his wife, but considered her trivial problems to be amusing sometimes and annoying at others.

"Well, Mr. Sontag, I'm sorry Mrs. Sontag is in such a dither at my arrival. I'm really not hard to please, and I will be appreciative of anything you do. I'm looking forward to our meeting."

"It is not far from here, Mr. Daly, and it's a nice ride through the country."

Before long, they entered the Sontag Farms. The Belgians were out in the pasture. Even the colts looked huge. They were such large creatures and yet so gentle. They were slow moving and ponderous but sometimes the colts would give a little buck and run races with each other to remind the viewer of their young age.

The door opened as the carriage arrived, and Mrs. Sontag came out to meet them. "Mr. Daly, we've been waiting for your visit. I can't believe someone from America is here. I do wish Mrs. Daly could have come, but you will tell her that the next visit she must. Now do come in and we will have tea and you can rest before dinner. Or do you want to take him out to the horses? Oh, Mr. Sontag, I knew you would want to rush out to the barns. No, I want you to have tea and then if you must, go after tea. Poor Mr. Daly! Don't let my husband wear you down. He loves his horses and spends most of the day working with the horses and veterinarians or on books and records of the horses. I can barely get him to take time away." Marcus had no time to answer, nor did Mr. Sontag, for Mrs. Sontag asked and answered her own questions. There was no doubt that Mrs. Sontag was a handsome woman. She was tall, thin, blonde, blue-eyed and vivacious. Marcus noticed her hands, for Joan Sontag's long slim fingers were used expressively while in conversation. Her fingers fluttered softly as she gestured to make a point. It was hard to believe these gentle hands sometimes handled the Belgian giants. Marcus also was amazed at her English, for Mrs. Sontag and Mr. Sontag both spoke English fluently, as they had been schooled with English as their second language.

Marcus said, "Mrs. Sontag, tea will be most welcome, and then, like your husband, I would like to go see the horses. We men who deal in horses find nothing more interesting than those animals. I, too, spend too much time on my work and my animals, anyhow."

Mrs. Sontag prepared the tea and poured. She had selected the china with the burgundy border, although she would have preferred the flowers and bird china. The tea went well, with Mrs. Sontag carrying the better part of the conversation. Mr. Sontag would have his turn, and he found it relaxing to listen to her tales of "my day".

After tea, Marcus and Mr. Sontag went to the tack room. There on display were the records of the selective breeding that had resulted in prize-winning Belgian horses. Marcus was interested in the records and the scientific way Mr. Sontag had taken to arrive with the top horses of the breed. Marcus wanted a stallion and two mares, as he already had some Belgians at Riverside. With this addition of champion stock he would have the best of the Belgians to work the farms at Riverside. It didn't take Marcus and Donald Sontag long to come an agreement, and the deed was done.

That evening, Mrs. Sontag prepared a dinner that was delicious. She worried that Marcus might not like what she had selected, but he did and made many complimentary comments about the food and her hospitality. Mrs. Sontag said, "You know, everyone thinks the French have the finest cuisine, but actually it is the Belgians. Our food and preparation is the finest in Europe. It is lighter than German, but as robust in taste. It is as light as the French and as attractively served. Belgians do not boast about their food preparation, but we should."

Marcus said, "You are right, I'm sure, for I've never had a finer meal. The fresh fish course of sole was delightful, the roast beef superb, and such wonderful vegetables! I'm not really a vegetable man, but they were remarkable in taste. Now that we are through with dessert I can vouch for the delicious cake such as I have never tasted before."

Mrs. Sontag beamed, for she was proud that she had pleased this man from America and that at least he knew that Belgium had the finest food as well as the finest Belgian horses.

Up early the next morning at the Sontags, Marcus selected the horses, gave Donald the necessary information for shipping and made his payment to the Sontag Belgian Horse Farm. Marcus was back in Brussels by late afternoon and boarded the night train to Paris. He would have a berth and a good night's sleep before Paris and Margaret and the children.

The arrival of the train in Paris was at 8 a.m. as scheduled. The train came into the Paris station, with its arches of iron and its glassy iron roof. The light coming through the roof made a pattern on the train platforms and walk areas. People poured from the train and hastened to the drivers of the hackney carriages called "hacks". Everyone jockeyed for position so that they could get a driver's attention. Marcus joined the mob and made his way with the porter to the cab area. As he came forward to hail a hack, a carriage appeared from nowhere. The driver opened the door and Marcus saw Margaret. She had come to meet him. Marcus did not expect this and was amazed and amused. "Dear Margaret, me darling, you do care about your old Marcus, to get up and out at this time of day? My wife and companion, you are wonderful. Let me look at you. As always, you are beautiful and your smile makes all the world right for me."

At that Marcus embraced Margaret and gave her a kiss which she returned. It was not proper, but at 8 o'clock in the morning who is out watching one's manners?

The carriage took them to the apartment, which was located close to the Louvre Museum in a beautiful section of Paris and certainly had a "good" address. Marcus helped Margaret from the carriage, and the driver put the bags outside the carriage. The doorman at the building came forward, greeted Mrs. Daly and met Mr. Daly for the first time.

"It is my pleasure to meet you Mr. Daly. Welcome to Paris. Mrs. Daly and your children have settled in nicely and we are happy to have them and you in Paris. I am especially happy to have you in my building. My service is at your call." The doorman took the luggage and held the door for Mr. and Mrs. Daly to enter the building.

The older girls were at school and would not be home until Friday night. The young ones had not been dressed and readied for the day, but they were up and anxious to see their Father.

"Where is my Papa?" said Harriot. "He is to be here this morning, I want to go now, even in my night clothes. I heard the elevator."

The maid soothed Harriot and said, "Harriot, you want to look pretty for your father, and certainly you are rumpled from sleeping all night. You will be the most beautiful princess when he sees you."

That satisfied Harriot, for her father called her "Princess" and that always conjured up visions of beauty and happiness. And, yes, she wanted to look like his princess.

Marcus II was just getting up and although he was anxious to see his Father, he was not as vocal as Harriot. Marcus II had heard Harriot as he awoke.

Marcus said, "Margaret, where are the children? I know Madge and Molly are at school until Friday, but where are Marcus II and Harriot? I thought they would be here to greet me."

Margaret replied, "Marcus, the children are just getting up and will be down in a minute. You will know very soon how much they love and miss you. Let's go into breakfast. They'll come to the dining room as soon as they're dressed. We can have a cup of coffee and a glass of orange juice while we wait."

"First I need to wash my hands. Where is our bedroom so I can do so?"

"Oh, Marcus, I'm sorry. Come, I'll show you our room. I

love it, but I hope you don't think it's too feminine for you."

Marcus entered the room and discovered what Margaret meant. It was lavender and lace and ruffles and pillows. Marcus looked for a moment and then turned to Margaret and said, "Don't you ever tell anyone I not only stayed in this room, but also I actually like it. Margaret, it looks just like you and I love you, so it will be glorious to be with you in this soft, fluffy feminine room."

Margaret actually blushed, for the absence from Marcus had made him seem as if he were a stranger. She needed to get acquainted again, and here he was going to share her bed. She laughed at herself and told Marcus it was their secret from the world. None would know of the feminine boudoir in Paris that Marcus and Margaret shared.

Margaret went into the dining room and Marcus came shortly. They had just seated themselves when the children arrived and Harriot hurled herself at Marcus. "Papa, where have you been? Mama read your letter to us and I want to go to Ireland. When are you going back? I want to see the horses, too."

Marcus II was used to waiting until Harriot had her way, for he was a patient, quiet boy and never assumed that he should be receiving more attention.

Marcus turned, put his arm around Marcus II and said, "My lad, you are the man in the family with these women. Have they been good and caring of you and you of them?"

"Oh yes, Papa, but I can't carry on your part. You are what we want to be head of this family. I try to do what is expected, but I think I'm too young."

"Right you are, my boy. I shouldn't expect too much, but you are a fine boy and I'm proud of you. You are the man in the family when I am gone."

Harriot said, "Mama, why don't you go somewhere so I could be head of the family as the woman of the household?"

"Where does she get these notions, Marcus? I think you encourage her to be a free spirit. A lady is reserved, and Harriot is anything but reserved," said Margaret.

Harriot heard her mother, but chose to ignore it and just held herself closer to Marcus, her dear Papa.

Breakfast was served, and the children ate well, listening to Mama and Papa and sometimes breaking into the conversation with some bit of information most important to them. Marcus loved it. It was easy for his life to be full of projects not involving his family, but when he was with them he realized how lonely he was for the warmth of his home. Nevertheless, Marcus was with his family and he was going to enjoy all that he had missed while away from them.

The children left for school. The maid cleared the table and Margaret took Marcus' hand and took him to see the apartment. As was usual with French decor, there were mirrors everywhere to give the illusion of spaciousness, for the rooms were not as large as they seemed. The furniture was white with gilt detail and upholstered in grey brocade with embroidered pastel flowers. Marcus nodded yes to what Margaret asked and he did like the apartment, but somehow he did not feel that there was anything of him there except his picture. That was all right. Paris wasn't his home, and he really was a visitor. If it pleased Margaret, it pleased him. They ended the tour in the bedroom and sitting room. They had much to say to each other in the privacy of their rooms. Distance makes the heart grow fonder and they reaffirmed their love and respect for each other.

Marcus was listening to Margaret's soft breathing. As she slept, he turned to look at her, thinking, Margaret Evans Daly, my wife. Who is she? What is she? Remembering the young woman he married, he thought of her brown hair soft to touch, her smile which came from inside her and spread genuine happiness and love from her soul to his. It melted his heart. Could he love her more?

Then there was Margaret the hard-working wife who was not afraid to work, who even took in washing so that she could help him through their financial crisis while he tried to find operating expenses for the Butte mine. Still there was the smile and love embracing him. Yes, he loved her more.

Margaret desired to guide the children through the pitfalls of other heiresses who were wed to men who desired their money instead of them. This was a hurtful period, for the children came first and Margaret with love and duty often forced issues. The outcome had been best for the children. Still, this has been a hard time for me, he thought. I have a difficulty being second, and with Margaret that's the way it must be. Do I love her more? Well, I don't love her less. I understand and Margaret tells me we will have our time together soon. I wonder . . . but, Margaret, I pledge ye my troth and I would repeat the vows again for, of all, you have given the most to create love and kindness in the family. Maybe I do love you more — I know that I long for you when we are apart. I could not go on without you. Whether we are together twenty-four hours or separated by circumstances, you are the one, Margaret.

Marcus' thoughts continued on and, as he turned toward the wall, he heard Margaret sigh and her arm came across him.

Margaret holding Marcus as he slept as was their way.

The next day, Madge and Molly would be home for the weekend. Marcus wondered where the years went. Madge was seventeen and Molly, fourteen. Certainly, Madge was a young lady and Molly not far behind.

Friday afternoon at 3:00 their carriage stopped at the front of the building.

"I can hardly wait to hear about Papa's trip to Ireland," Molly said.

"I want to hear about London," said Madge, "We've had English and French history in school, and I know I would love London. I do like the cities, and Ireland is so rural."

"Better not let Papa hear you say that, Madge. He is proud to be an Irishman, even if it is a social burden sometimes."

Molly and Madge went directly to Marcus and stood at his side. Each gave him a kiss on his cheek. Marcus grinned, "Girls, let me look at you. You are beautiful young women. I find it hard to believe that you are no longer children. You are adults."

Madge had a navy blue and white dress with a bustle, while Molly was dressed in less formal fashion in a dark green dress with ecru lace at the neck of the jacket. Madge shoes were black and Molly's brown. The were most attractive and well groomed. Their brown hair was tied back with ribbons and their blue eyes shone with happiness because they were a family again.

Madge was the quiet one and was her mother's best friend. Molly's eyes sparkled with happiness and interest in everything around her. She liked and respected Madge, but they did not share many interests. They each had different friends and enjoyed different activities. Madge preferred quiet entertainment in her spare time such as needlework with her mother, reading classical books and looking at fashion magazines. Molly was interested in the world and read the newspaper daily to glean news of the world. She knew that she was proud of being an American, but she also enjoyed studying other cultures and giving them credit for their contributions to civilization. She loved tennis and even roller skated. Of course they both loved to ride their horses, as did all of the Daly family. Activities were more satisfying to Molly when they were exhausting. Madge tired easily, and Molly was always full of energy. These two children were quite opposite.

Both Madge and Molly thought of Marcus II and Harriot as their dear little ones, for the age difference was nine years to eleven years. Marcus and Margaret really had two families. As a result the older girls aided and comforted the two younger ones. They doted on Marcus II. As the only boy, Marcus II had an exalted place in the girls' hearts. Marcus II was like Madge, a quiet young man and really quite serious and studious. He was eight years old and his tutors found him a quick student. He looked like his mother's side of the family. He was tall for his age, thin featured with straight brown hair. Like Madge, he was quiet and loved his books. He had become quite interested in zoology. His greatest pleasure was to go into the country and walk with a companion or ride with his mother and sisters through the bridle paths, identifying the birds. That was what made Riverside so special to this young boy who was old beyond his years.

Harriot was full of energy and busy, busy, busy. Horses were her love, and at six years of age she insisted on a horse for herself in Paris. Harriots horse was at a stable by the park. The rest rode leased mounts kept for such use at the stable.

"Papa," asked Madge, "did you like London? I know I would."

"Yes, Madge, I did like London. My visit was hurried but I did walk about some and I had the carriage drive me to some of the landmarks. I would like very much to return."

"Margaret, why don't you take the girls to London during the spring recess from school? Leave the younger ones here in the capable hands of the housekeeper and have a holiday. London is a great city."

"The horse farm is only fourteen miles from London and Mr. Lancaster would be glad for you to visit. If you decide to go, I will have him meet you, get you reservations in London for opera and the theater, as well as arrange for your hotel. Then you could go to Epson to see the farm and get to look at a bit of the country-side outside of London."

"That does appeal to me, Marcus. How about you two?" asked Margaret.

"We would love it. At least, I would love it," said Madge.

"You know I would like that better than anything," said Molly. "I've wanted to go for ages, and it will be good to hear English spoken all the time. My mouth is getting tired from trying to speak French without an accent."

Marcus laughed, "Well good luck with your accent, darling. I'm still speaking English with an Irish brogue and it has been many a year. I was just a year older than you, Molly, when I came to America."

There was a great commotion in the hall and Madge said, "Here comes Harriot. She probably has a stick and a ball, playing polo in her imagination."

And that was right for Harriot did have the stick and ball. She picked up the ball and laid it and the stick in the corner. She looked at her mother and saw a firm look, then she looked at her father, who had a twinkle in his eye. Harriot ran to her father and sat down beside him in the same chair. Marcus hoped the French furniture was stronger than it looked, for Harriot sat down with a flounce.

"Papa," said Harriot, "I arranged for my own piano lessons. I told the governess that Mama said I could learn to play the piano so she got a teacher to come. Mama said I told a fib for she hadn't said I could have piano lessons. She just doesn't remember, Papa."

Margaret was not pleased at Harriot's actions nor words. "Harriot, we have settled the piano lessons, and you are going to get lessons starting next week after your father leaves. We will spare him that. And dear, please try to be more lady-like. You really are a pretty girl, but your actions detract."

Harriot smiled and said, "Oh, Mama, I didn't know I could take lessons. Thank you, Mama."

All was well and the family continued to enjoy each other. The conversations became quite spirited at times. Harriot and Molly kept the conversation going while Madge and Marcus II tried to get their thoughts into words so that they, too, could converse with the rest.

Harriot was six years old, blonde, blue eyed, vivacious and assured. She was dressed in blue taffeta with a pink sash draped in swags across the front and a large bow in back. Her little legs had white stockings buttoned to her camisole and topped by white panties with an embroidered band of lace on the panties' edge. Her shoes were black with straps across the foot and ankle fastened by buttons. Her hair cascaded down below her shoulders and was caught in the back with a bow ribbon that was almost as big as her head. The sight of her melted Marcus' heart. She was his princess.

"Papa, did you get me a horse? I'm learning to ride and my instructor said I was a natural. What's a natural?"

"Harriot, darling, first I have a horse at Riverside waiting for your arrival this summer. You will name her, for I have not done so. A natural is someone that can accomplish a task without special instruction. It is instinct."

"Papa, what's instinct?"

"Margaret, take this question. I'm needing my brain for other things."

Margaret patiently explained "instinct" to Harriot and hoped her explanation would not introduce another question.

Tea was served and the children and grownups enjoyed the finger sandwiches and lemon cakes. After tea, Harriot and Marcus II went to play in the park with the governess and nanny. Molly and Madge went to their sitting room, which had a bedroom on each side, making a suite for the young women. They would finish their studies early so that the weekend would be free.

Harriot and Marcus had the same arrangement, but instead of a sitting room, it was a playroom with toys. The governess's bedroom was off the playroom. One end of the playroom served as a school room, and after school it was for recreation for the children. To keep Harriot's attention, their governess placed her desk in front of the bookcase. Marcus II and Harriot's desks faced the teachers, and during school hours the rule was "eyes to the front". The play area was behind Marcus II and Harriot. A blackboard was on a post-and-beam easel sturdy enough that even Harriot wouldn't knock it down in one of her exuberant play rides through the room.

Marcus and Margaret decided to take a carriage ride through the city to familiarize Marcus with some of the best-known landmarks. Since their apartment was at the east end of Champs Elysees they took the carriage to the Louvre, which was joined by the Jardin de Tuileries. The gardens were designed in French style with basins, fountains and many statues and were surrounded by terraces. Notre Dame Cathedral, with its flying buttresses, came into view. The carriage tour took the Dalys to the recently completed Eiffel Tower, for it was constructed for the 1889 Exposition, and it was a wonder of iron construction though the iron was really quite lacy looking. "I can't explain this tower," Margaret said. "I've never seen anything like it. Never mind, we will see what you think, Marcus," said Margaret.

When they came in sight of the tower, Marcus said, "What do they plan to do with this structure? Does it have any reason for being except as an ornament?"

"I don't know Marcus, I can't think of any use for that tower after the fair, but people seen to be fascinated by it. It has been visited by thousands of people."

"Perhaps they could use it as a wireless station," Marcus said.

"I understand they measure moisture in the atmosphere at different levels from the tower," said Margaret.

They rode back by the River Seine and returned to have dinner at their home. The children had eaten, even Molly and Madge, for they had been at school in the morning and had had a busy afternoon talking with the family and doing their studying.

The days went by and the weekend was almost over. Marcus had called his agent in France and he had a horse for Marcus to look at that could enhance his breeding stable. Marcus hated to take a day from the family, but he wanted to see the horse for himself.

He did so and was pleased with the horse and paid the high price that French breeders demand for their horses. The French are convinced that their selective breeding, stabling, care and training are superior to anyone else's and they charge accordingly. Marcus paid the price, but he was not unhappy for the lineage, and the horses were impressive.

It was wonderful for Marcus to be with Margaret and the children. It also made him realize that his life was very lonely despite his busy schedule. However, he had his projects, and it was necessary for him to be away from his family. The children had to have the finest schooling, for they would have enough money and position to take important places in the world.

After his time with the family, Marcus boarded the train to the Port of Le Harve to begin his trip back to New York. He had told Margaret he wanted to go to Ascot in June, but it wouldn't be this year. The mail Marcus had received in Paris and London indicated that he was needed in New York, Butte, Anaconda and Riverside, all at the same time. Marcus's farewell with Margaret was tearful, and the younger children wanted to come with Marcus back to the U.S. Marcus calmed them all.

"My dear family, the time will soon be here for you to come to Riverside, and I promise to be with you all summer. You go to London and enjoy the tour and study hard, and then it will be summer! I will send you letters, and you must write to me, for I'm alone, and you have each other."

"Oh, Papa, are you all alone?" said Harriot. "I'll come with you, and then you won't be lonely."

Marcus patted her head and said, "Thank you, my dear, but your mother needs you here. I'll manage, but I do miss all of you every single day."

The trip back to New York was uneventful. Like his voyage over, on his return trip he met and talked to people; and, as before, he ate too much at the lavish dinners served six times a day.

The voyage gave Marcus time to read and digest his many packets of mail. He had to decide what to do about problems with Anaconda Copper, the lumber business in Hamilton, the construction of his race track and plans for horse training. One thing Marcus decided was that the horses needed year-round training, so a winter race track would have to be covered and heated.

On this tour of Europe, England and Ireland, he had learned some very important lessons on horse breeding and training: Heredity and blood counted most as a foundation; the dam and the colt were given the very best environment and the stallions separated but equally well treated; the training was begun early and was continuous; after the training came experience in races and tests; and the most skilled leadership was required, so the trainer and his staff riders and jockeys had to be skilled and experienced. Marcus had seen the results of this selective breeding, environment and training, and he knew he had seen the finest.

Riverside would be the perfect environment. He had the best trainer, Sam Lucas, and he would make the barns and the tracks conducive to the best results. Then he would build race tracks at Anaconda and Butte for the tests of his thoroughbreds.

The mine workers would like the recreational pastime, and the wealthy who have always loved the horses certainly would come to the races. Marcus knew that this venture would be a success and it was a labor of love.

The letter from his editor at the *Anaconda Standard* wanted to know if he should write an editorial defending Marcus against William Clark's paper, which had been maligning Marcus' good name. This had all started when Clark lost his bid for Senator of Montana Territory in the election of 1888. He blamed Marcus for that. Since that time Clark had bribed people to say that Marcus bought votes and told his miners how to vote. The paper also accused Marcus and Anaconda Company with illegally cutting timber on public lands. All of this was being printed daily and Marcus decided he should defend himself. He told his editor to use only facts, but to retaliate against William Clark. He thought to himself, "If I see him I'll punch him in the nose, that conceited, conniving hypocrite!" And Marcus would do just that. With that note written, Marcus vowed William Clark would not hold a public office as long as he had any breath left in him to fend off Clark's political ambitions.

Since Marcus could not receive any mail after the ship had sailed, the remainder of the voyage was enjoyable and Marcus had a good rest.

Landing in New York, Marcus went to the Nederlands Hotel and unpacked for a brief stay in New York.

Marcus took out his record books and started planning his itinerary for the rest of the winter and spring. He planned to spend the summer at Riverside. He would take care of the mine in Butte and the smelter and assign responsibility for the production to his managers. He could always be reached by telephone and could arrive in Butte or Anaconda in a day. The summer would be for building his stable of thoroughbreds. Already the stable and barns were ready, and some of the horses purchased at the races were stabled there. The European stock would start to arrive in the summer.

Marcus spent two weeks in New York. He met with the Rockefeller group, for they were interested in mining and had approached each partner in Anaconda Copper Company. It looked as if everyone wanted to sell except Marcus, but right now they would go with Daly's wish. How long Tevis and Hearst would stay, he couldn't tell. Marcus thought Haggin would support him always.

Marcus went to Anaconda and Butte in the spring and took stock of what was going on in Butte. Marcus went to his office and asked the mine manager to report to him at 2:00 that afternoon. At 2:00 Mr. Digan came for his appointment.

"Marcus," said Mr. Digan, "it's good to have you back. The mine is producing good ore, and we are shipping to the smelter as fast as we mine."

"I know that, Joe, but I want to know where are all these new companies are coming from? I understand there is a young man named Augustus Heinze who's the talk of the town."

"Well, Marcus, Mr. Heinze is an educated young man from the east. He doesn't have any family, but he does have connections with Eastern money. His first venture here was with Jim Murray. You know old Jim is pretty shrewd. Well, that Heinze got the best of him. He leased the mine from Murray, and when that lease ran out he offered Murray a higher percent (50%) on 15% ore and nothing on 2nd class ore under 15%. It sounded good because all the ore had been 15%, but as soon as it was a signed document the ore never was 15% again. Heinze's miners just caved down the country rock beside the vein and mixed it with the ore to be sent to the smelter. Murray sued, but it never got him anything. Heinze had a judge that favored him when there were law suits. You had better watch him, Mr. Daly. He is a slick one, and all the women like him, too. You know -- good manners, good dresser, good looking. I mean, he is something!"

"Thanks for the warning, Joe. I'll try not to get involved with that one. How about Clark? Is he with Heinze?"

"No, he doesn't get involved with Heinze -- doesn't give him a chance. You know Clark, he hangs back and lets others do battle. Then, with his luck, he picks up the pieces."

"Well, I'm going to do battle with Clark. He wants Helena for the capitol. Hell, there aren't enough people in Helena to have the State capitol. The population is here in Butte and Anaconda, and the main businesses of ranching and mining are here. Why does he want Helena? I already know, Joe. It's because Anaconda is the Anaconda Company town. But it is the right place to have the capitol. I'll battle for Anaconda."

"I know you will, Mr. Daly, and I can tell you, you don't have to tell your workers how to vote. They want Anaconda as much as you do. In fact, the miners from Clark mines and ours are having some brawls right now, and Anaconda and Helena are causing it all."

"Well, enough, Joe. I'll do my best to get our capitol at Anaconda. Thanks for the information. You keep up the good work and keep the miners happy. By the way, tell them there is going to be a thoroughbred race track in Butte and Anaconda, and on opening day the miners will get a day off with pay."

"Whew, Mr. Daly, that ought to make them happy to be with Anaconda Copper Company. Wait until they tell the miners in the other copper mines."

"All right, Joe, let's hope the price of copper stays up and steady. We've a lot to do."

"Goodbye, Mr. Daly. It's good to have you back."

"Goodbye, Joe."

Of course, Helena became the capitol of Montana, and Marcus's beloved Anaconda lost. He had good support among his friends and workers, but the opposition was able to gather more support. Politics at that time was tough, and the winner depended on the person who could gather the most votes for his cause or his candidate, even if it meant buying votes with money or favors. There were fights and accusations, but Helena became the capitol. Rumors started that twice as many votes were cast by citizens of Helena as there was population. The ballots had been counted in the sheriff's house in Helena. Marcus was very unhappy & voiced his anger to all who were within ear shot. Years later a secret room was discovered with the ballots intact. That was when the double voting deed was confirmed.

Chapter Eight

In March of 1890, Margaret Daly and the two older girls drove up to the Savoy Hotel on the Strand in London. The theaters were in that area, and one of the pleasures they looked forward to was seeing the plays. The opera was in the market district not far from their hotel. All three were excited and very happy to be in London at last.

The doorman of the Savoy knew who was to arrive from the boat train. It surprised Margaret that he welcomed her by name. The desk clerk also knew who she was, which made her feel welcome and at home. The lobby had marble floors and marble pillars and it was very grand. Large crystal chandeliers glistened with the new electric lights. The hotel certainly had modernized quickly, for all the rooms had electricity. It was exciting, for they had become used to the smaller scale of the French buildings. Of course, George V Hotel in Paris had this same large-scale grandeur.

Mr. Lancaster had made all the arrangements for Mrs. Daly and her daughters. He left a note in the box and explained to Margaret that he was sorry not to have met them, but he had arranged for their arrival and hoped all was satisfactory. The theater tickets were for two plays the next two days, then an opera and another play. He was especially pleased that "Turnadot" would be there at this time. "I understand," he wrote, "that it is the most lavish setting of all the operas. Aida may need a larger stage setting, but Turnadot is rich with brilliant silks and brocades of gold. It is a Chinese story." Margaret told the girls the plans Mr. Lancaster had made. The four days were filled with good food, good entertainment and the pleasure of the city of London.

Margaret wrote Marcus a letter about their stay.

"Dearest Marcus,

You were right to encourage us to come to London. We've been so busy, so delighted and tired from excitement. The stage plays were excellent and, of course, here Shakespeare is done as it should be! The opera was all they said it would be. I understand this opera's Turnadot is not performed as much as others because it is so expensive to produce. I believe it, for the costumes and properties were sumptuous!

The girls have loved hearing English, and I hear them practicing the British accent. They have had such a wonderful, exciting visit and will be able to settle down and work hard on their studies (I exacted that promise before we came on our trip).

Dear Marcus, we do miss you, and I'm particularly sorry that Anaconda lost to Helena as the state capitol. Do not belabor that, Marcus. You did all you could. Please accept it.

It won't be too long before we will be together at Riverside. We will sail from Le Harve the first week in May and will stay in New York just long enough to organize ourselves for the trip.

Do you think you could be in New York to meet us? I would so like you to do that for me. Then we can journey to the Bitterroot together. I love the train ride across the United states. It is always beautiful and interesting and always will be no matter how often we travel that territory.

We are leaving tomorrow and will be in Paris soon. We all talk about you every day and Harriot doesn't understand why her Papa has to be away to do his work!

My love is yours, Marcus. I think of you always.

Your loving wife,

Margaret"

Marcus received her letter in three weeks and he relived her stay in London through the pages. Margaret's love always came through the pages. Such a woman, Marcus thought, and she is my wife and my fortune.

It was May, and Marcus was still in Butte and Anaconda trying to keep everything together and ignore the distractions. The race tracks would be built that summer and by the next season there would be a full schedule of racing. Mr. Haggin had a stable in California and planned to bring some horses to Butte and Anaconda to race. Of course, Marcus would have his horses ready, and it was open to anyone with horses that could compete. They might be able to have some of the races at the end of the summer. It would be a trial to see it all was done properly. Marcus called for Mr. Olson who had built for him on other projects.

"Ollie, I have plans for a racetrack and stadium. The specifications are exact and I want them followed precisely. I will not be here all the time nor will I be able to supervise, but I feel you are qualified. I want you to have the contract. I want an estimate, but it does not have to be exact. I'm depending on your fairness, for I have no other contractor in mind."

"Yes, Mr. Daly, I would like to do this and it will be a great honor for me. I love the horses myself and will like being a part of the sport. You want a track in Anaconda and another in Butte?"

"That's right, Ollie. We'll fill these stands with horse lovers and gamblers, for it will be a great sport for everyone to enjoy," said Marcus. Ollie looked at the plans and said he would go to see the locations. He would have the estimates in three days.

"Fair enough. I'll see you in three days," Marcus said.

In three days Ollie was back with the estimates. They seemed fair to Marcus and he signed the contract, making note that the tracks and stands would be ready by August. August would be a good time for the first races, as the weather was as predictable then as anytime in Montana. Marcus wanted dry weather for the races, and August was the best chance. The track was to be exact for the races and built just as the tracks in New York and California, with the distances precise.

Since Margaret's letter Marcus decided he had better meet her and help her make the journey to the Bitterroot and Riverside. It was still a hard trip although the private cars did make it easier as they were switched to other lines without loading and unloading their belongings. Margaret would be getting to New York the middle of May, Marcus would be there the week after, and then the packing would start for the summer at Riverside. Marcus would go to his Maryland horse farm. He intended to race at Belmont as soon as possible. The races in Butte and Anaconda would be good training for his horses.

While Marcus was riding on the train, he decided a lot of his life was saying "hello" and "goodbye" to his family. At least at Riverside they would be under one roof, including Margaret's mother and father, sisters and brother-in-laws. It went well the past summers, with the men agreeing it was like a retreat to be away from crowds, noise and business. The women liked it because they were together for a good visit in their own world, and the children were happy to be free with so many different activities.

After going to the Nederland Hotel, Marcus sent a messenger with a note for Margaret.

"Dearest Margaret,

I arrived on the 3:00 train and decided not to upset the household at dinner, so I will be over at 10:00 when the children are asleep. Even the girls should have retired. I need some time to be alone with you and would like our privacy for a while.

Send word back with this messenger. I hope it is not too late or complicated for me to come over tonight. I've told him to wait for your answer. Anxiously awaiting your reply,
Your faithful one,
Marcus"

The messenger returned with a reply from Margaret. It was short and written hurriedly. Margaret didn't take the time to write in her precise rounded, most legible handwriting.

"Dear Marcus,
Do come. I will be waiting. The children and the servants except for Cutler will be in their suites or rooms and hopefully asleep.
Love to you, dear Marcus,

Margaret"

That evening at 10:00 the carriage stopped at 725 Fifth Avenue and Marcus Daly entered the building. The faithful Cutler was at the lower level to take Mr. Daly on the elevator to Mrs. Daly's suite.

"Hello, Mr. Daly, it is good to have you back in New York."

"We won't be here for long, though. We are going to Riverside soon," said Marcus in greeting Cutler.

"Yes, sir, we are already busy packing and arranging for Riverside."

The gold and glass elevator cage stopped at the second floor level and Marcus made his way to Margaret's suite.

They embraced freely, for they were alone and were able to show their emotion.

Marcus said, "Margaret, I could cry you are so beautiful and lovely. It all hits me hard when I'm with you, and I wonder why we're ever apart."

"I know, Marcus, but it'll be a few years more of your work and then you can retire. The children will be grown, and you and I will have our time together. I dream of it all the time," said Margaret.

Marcus took Margaret's arm and guided her to their bedroom where the canopied bed curtains would close out the world and become their private sanctuary. Those who have had true love will know that the sacrifices they have made for each other only deepens the mystery, beauty and holiness of their marriage bed.

The next morning when the young people were at breakfast, Margaret and Marcus came into the morning room. The greetings were exuberant, and everyone was talking at once. Through all the noise each child seemed to be able to convey their message to either Margaret or Marcus. Marcus smiled. "Come, let us calm down. Your breakfast will be cold, and your Mother and I are going to be with you for the next four months. We will be going to Riverside soon, where we'll have the whole summer together. I want to show you the progress at Riverside and the farms. Hamilton is getting to be a very nice town, and I just know you'll enjoy going into the company store. There's so much for you to see and do that this may be the busiest summer you'll ever have.

"In August we'll go to your Grandparents' house in Anaconda and to the horse races in Butte and Anaconda. There you will see the Daly colors on the jockeys. They are your colors, too, you know, not just mine."

Madge asked, "Papa, are there racetracks in both towns?"

"Yes. They will be finished by August, and horses from California and some trotters from the midwest will be coming. We will have both sulky and horse racing -- not on the same day, but each will have their separate time."

The announcements about horses heightened the desire of everyone to hurry on their way to Riverside. Harriot asked, "Mama, can I go get my packing done so that I will be ready to go today?"

Margaret said, "Harriot, you certainly can start packing, but we are not going to leave today. But, it is a good idea to get your packing done early. Then, if you forget something you will have time to add it to your treasures. Mrs. Cutler said she would help you and Marcus II get ready to leave for Riverside. You might ask her, Harriot, when she will have time."

Harriot said, "I'll go right now and find Mrs. Cutler." The last words were lost as Harriot was out of the room before the sentence was finished.

Marcus II said, "Papa, will you come with me? I want to show you my fishing gear. Mr. Cutler went with me to Abercrombie and Fitch and helped me select my fishing pole, line and tied flies. I love to go fishing at Riverside. Mr. Cutler said he would take me to the Bitterroot River this year. He thinks I am old enough to river fish now."

Marcus was pleased, because Marcus II was a quiet boy and didn't need much attention. Marcus said, "Good, my boy, I'm pleased you like to fish. Actually, I'm not a very good fisherman, but I won the award for the largest fish from Lough Corrib in

Ireland. I was the most surprised fisherman there. Let me see what you have, Marcus."

Off they went, and Marcus reached down for Marcus II hand. Marcus said, "Marcus II, we can't keep calling you that. Since I am Marcus, would you like to be called Mark?"

"Oh, I would, Papa. I'm not a Junior, and Marcus two sounds funny. I would be pleased to be called Mark."

"All right, as of now you will be called 'Mark' and we will tell the household tonight at dinner."

Mark was pleased and his smile was broad. He and his father had a good visit and one could hear Marcus remarking about an item and Mark explaining and then laughter between the two. It was a good time for father and son.

"Papa, did you see the flies I have? They are such pretty colors. He showed his father and Marcus said, "This yellow, black and red fly is a beauty. It is a masterpiece tied by an expert."

"Yes, Papa, that is Mr. Conkle's fly. He ties for Abercrombie and Fitch."

In New York, they looked at the fishing rods and a B.B. gun for target practice. Mark also went into his wardrobe that had four drawers and a cupboard, with two doors above. In one of the drawers he had an animal book, realistically illustrated, that told of the African wild animals.

"Papa," Mark said, "this is my favorite book. I've read all about these animals and their natural habitat. I hope to go to Africa someday on a safari."

"Mark, I'm sure you will be able to if you learn to be a sharp shooter. The animals are wild and self preservation is the strongest of instincts. They attack if they feel threatened. A quick and accurate shot is required for the hunter to survive."

That night at dinner the household was told that Marcus II was to be called Mark henceforth. Everyone agreed that was a much better name than Marcus II.

Marcus made a quick trip to Maryland to look at his horse farm. The horses would come by rail from Riverside to the Maryland horse farm. They would train and rest and acclimate the horses before they raced at Belmont Park during the racing season.

Good managers for the farms were essential. Marcus could not personally oversee everything, and selected managers wisely. The managers and the workers were rewarded monetarily as well as with praise and confidence from their employer. Pleased with the Maryland location of the farm outside of Baltimore and the manager, Marcus returned to New York and made ready to start the journey to Riverside. Before leaving New York he received notification that Mr. Lancaster had sent breeder stock to Honoria and her farm manager, Mr. Digan. The animals sent were the finest sheep, pigs and cows to strengthen the stock. A letter from Honoria was in the morning mail.

"Dear Marcus,

Mr. Digan is raving to the neighbors about our new breeding stock. He also says they can have some of the results as soon as they reproduce. He knew you wanted the neighbors to benefit too. We will have the finest stock in Ireland. Thank you. Marcus, we do appreciate your kindness. We say a prayer of thanks each day for your long life.

I went into Cavan Thursday. I have not been in, except for church since last year. It was good to walk along the streets and see the people.

Tell your children and wife they, too, are in my daily prayers.

When you have time, write me a letter. I'm waiting to know how your life is.

Your loving sister,
Honoria"

Marcus read the letter and vowed to have a note to Honoria in the mail once a month.

At last the Daly household was ready to start the westward journey to Riverside. The Daly family and their support system filled the coaches. The driver drove to the private car area of Grand Central Station in New York. The Dalys had three private cars and a baggage car filled with family, servants and belongings. As the train sped through the countryside, Margaret said to Molly and Madge, "Isn't it amazing that our trip to Riverside is over half-way across the United States and we will be there in only four days? The invention of the steam engine has made our travel so convenient compared to that of our forefathers, who had the tedious and dangerous travel over roadless country, taking months to make this same journey."

"Mama," said Madge, "I never thought about that. I guess I should, for both Molly and I came in the stage across the mountains when we moved to Butte. I remember the dust and the bumpy ride very well. Do you remember, Molly?"

"Yes, I do, even though I was so young. It must have been very impressive to me. I remember I didn't like it, but, Mama and Papa were there so Madge and I felt safe." Very much on schedule, the Dalys and their entourage arrived at Riverside.

The servants quickly put the house in order. Mr. Cutler

said, "I'm glad we had the summer employees open the house. Otherwise we would have had to come before the Dalys and it is easier for us to all arrive together."

"I agree," said Stena. "That first year when we came and set up the household without the summer help was very difficult. We were tired before the work started for the summer. I really look forward to our summer workers. They are a good lot, these Bitterrooters."

The family had established their rooms and their routines on the first visit to Riverside. Now all was the same and each one was anxious to pursue their interests. At the end of the week the grandparents, Mr. and Mrs. Evans, and Mattie would arrive; the Lewis family would come in from California later in the month. Friends and family from Butte, Anaconda and elsewhere would keep the Daly mansion full of friendship and happiness throughout the summer.

"Margaret," said Marcus, "I want you and the children to come see the horses I've selected for you. They are stabled by the carriage house. Let's walk over there now. I had them make a path for us to walk on and the horses will use it when the stable boys bring them for you to ride. Also, I want you to post a schedule for the stable boys when they are to have the horses ready at the bottom of the steps behind the mansion. I assume you want to ride every day, Margaret. You are an excellent horsewoman. I don't mind telling you, you are never more beautiful than when you come riding up to me with your hair creeping out under your hat and face flushed from the run. I'm telling you, Margaret, you are an exciting woman."

"Marcus, don't say that. I'm not exciting!" Margaret's face

flushed and she lowered her eyes.

"Yes, you are, me darlin'."

"Marcus, for heaven's sake, let go see the horses. I'll get the children. Goodness."

The group gathered and went down the steps to the path leading to the riding stables. Harriot was shown her horse first. "Papa, is it really mine?" said Harriot.

"Yes, my darlin', this is your own horse. She will be yours to grow up with. You must learn the right way to handle her, for she has a soft mouth and will respond to your touch."

"What will I name her? I've been trying to think of a name." said Harriot.

"She is from Ireland and she has a little white spot on her forehead that looks like a shamrock — maybe you would like that name."

"What's a shamrock, Papa?"

"It's like a big clover that you see in the field. A bit of the magic of Ireland makes it a special and a good luck charm."

"Oh, then I will call her Shamrock. Isn't she beautiful. She likes me, Papa. She spoke to me -- hear that soft sound? I love her." With that, Harriot stepped forward and Shamrock leaned her head down for Harriot to pat. They would grow up together.

The rest of the group went to their horses. Everyone was pleased with the horses that Marcus had chosen for them. He had given a great deal of thought to the temperament of the horses and their would-be masters.

"Marcus, you outdid yourself. I have never seen more beautiful horses. We all will be proud to ride them and I, myself, can't wait. I intend to ride this afternoon." said Margaret.

"I'll ride too, Mama," said Mark. Molly and Madge agreed it was a good idea, and Harriot said, "Me too. Me too!"

"Marcus, are you going to ride with us?"

"Well, I was going into Hamilton to see Jim Hamilton, but I will do that tomorrow. You know the sawmill is built and production is at full capacity. You might like to go with me, Margaret."

"All right, Marcus, you ride with us today. I'll go see your sawmill tomorrow," said Margaret.

The Daly family gathered at the back garden steps and watched as the stable boys brought their prize horses to them. Mr. Cutler was in the kitchen at the big window, watching the family. He called Mrs. Cutler, "Come, look -- the whole family is going. I've never seen them all go together.

"My, don't they look grand in their riding clothes," he continued. "Mrs. Daly looks at ease on her sidesaddle. You can tell Miss Madge is not the horsewoman her mother is, and Miss Molly looks quite serious. Oh look, Miss Molly's horse is being fancy — maybe a little test there. But look at Miss Harriot, 'the queen of all'. Isn't she like a doll on that horse? There she is right up by her father. And Mr. Daly looks so proud to have his family around him.

"Where is Master Mark? There he is, just getting on his horse. My, he is handsome in his riding habit. I wonder what the Bitterrooters will think when they see this group in their Eastern riding habits? I guess they'll know that they are the Dalys. Then again, I doubt if any Bitterrooters will see them, for the Riverside estate has 22,000 acres. If they do see anyone, it will be the workers on the farms of Riverside," said Cutler.

The riders left two by two: Mark was beside his mother, Harriot with Marcus, and Madge with Molly. Marcus led the riders.

He chose to ride from the mansion over to the area that was once "The Big Corral" with nine square miles fenced for cattle. It was part of the Riverside estate, but there were some outbuildings there and a nice pond with trees at its edge. They could dismount and watch the mallard ducks and geese on the pond and walk around a bit before they returned.

"My, look at the mountains. There are the Como peaks above Lake Como. Isn't that a breathtaking vista? Doesn't it remind you of the Alps? Lake Como was named by Father Ravalli because it reminded him of his former home on Lake Como in Italy. Children, are you listening? Father Ravalli also had the first crops and domestic animals in the valley. He was a spiritual guide, carpenter, artist, administrator and farmer. The valley owes a great deal to Father Ravalli."

Mark called to his father, "Papa, is that the race track over past the first pasture?"

"Yes, it is. We'll ride down to it another time. It should be finished soon. The horses will be trained on the track, then we'll have some horse races."

"There is another race track that will be covered and heated for winter training behind the stock farm. Racing horses need good breeding, lineage, daily training routines, good environment, good care and nutrition, precise equipment and racing tracks standardized in shape and hardness.

"The Daly stable is going to have all of these prerequisites -- the finest available."

As they rode on, Molly said, "Papa, is that a barn on top of the hill further south?"

"No, that is an animal infirmary. We have a veterinarian in

residence and he will care for the horses and recommend treatment for other animals. But, the clinic building you see is for treatment of injured or sick racing stock. Dr. Knowles is the veterinarian. You all will meet him this summer."

The return ride from the pond area took them farther east, and they came to the front of the mansion.

"Mama, isn't Riverside pretty? I like the green trim and the beige paint," said Molly.

Before Margaret could answer, Marcus said, "Our racing colors are green and gold, but your Mama wouldn't let me paint the house gold." They all laughed. "Let's ride over to the greenhouse. Mr. Hughes should be working there this time of day. We can walk back to the house from there and I'll have the stable boys sent to the greenhouse for the horses."

Mr. Hughes heard the horses snorting and the chatter of the Daly family. He hurriedly put down his trowel and wiped his hands on a cloth he carried in his back pocket and came out of the greenhouse.

"Mr. Daly and family. I'm proud that you came to see the greenhouse. We have lots of plants in the hothouse beds. Are you coming?"

"Certainly, James. I want my family to understand what happens here at Riverside on all seven farms. What do you have inside?" asked Marcus.

Marcus introduced his family to James Hughes and they all followed single file behind Mr. Hughes and Marcus.

"Here, sir, we have 10,000 strawberry plants, 20,000 celery plants, 225,000 tomatoes, 12,000 cauliflower plants, and 10,000 peppers of all sorts." The children were wide-eyed at this huge expanse of little pots on the table and the number of plants.

"Mr. Hughes, " said Molly, "what will we do with all those vegetables?"

"Well, Miss Daly, what we don't use on the farms for food, we will sell to the town or ship them to Butte and Anaconda. Isn't that right, Mr. Daly?"

"Absolutely, Jim. the fresh produce plus fresh fruits are needed by the local people here and by other towns. With the railroad, we can ship the perishable foods quickly and retain their freshness."

Margaret and the young people said goodbye to Mr. Hughes and thanked him for the tour. The stable boys came for the horses and led them to the stables.

"Bye, Shamrock," called Harriot. "I'll be over tomorrow and we'll ride again."

"Now, Harriot, you probably won't ride tomorrow; but if not, I'll have a stable boy lead you around the stock farm," said Marcus.

Harriot's eyes showed disappointment at not going for a ride, but brightened when Marcus said she could at least be aboard and go for a walk riding her Shamrock.

Marcus told the others he would be along soon, after he talked to Jim Hughes longer.

"Jim, I think you are going to show a profit for the vegetable area. So far the farms are doing well. The only thing that won't be profitable are Mrs. Daly's flowers and her favorite flower, the rose. Guess you can make that up on the vegetables, eh, Jim?"

"I guess so, Mr. Daly. Mrs. Daly's flowers are the finest. I suppose we could sell some stock roses commercially."

"Jim, I was just teasing you. Keep Mrs. Daly's roses for Mrs. Daly. Whatever you develop in that line is for enjoyment, not for profit. You just keep up the good work. Good to see you. Let me know if you need anything."

"Thanks, Mr. Daly, we're fine here," said Jim.

Marcus left the greenhouse and looked toward the Sapphire mountain range. They had taken on a pastel blue with pink tinges as the sun moved farther to the west. It must be about 5:00, thought Marcus. The mountain views are best at that hour. The shadows cast by the different elevations gave a new perspective on the Sapphires. During the day they were flat and soft, and in the late afternoon rounded elevations at different depths showed the interesting contours of the Sapphire range.

After tea and dinner the children retired to the upper living room. Margaret and Marcus went into the living room with the other guests. The evening was a time to tell each other about their activities since their last visit. It was a lively conversation, for each had some event in their life to share. Marcus broke into the conversation and asked to be excused. He had some work in his office. He also wanted to read the newspaper and the daily mail from Butte and Anaconda.

First, the newspaper was to be read before the business mail. One item in particular made Marcus look closer and read carefully. A group of representatives from the Trade Union in New York had arrived in Butte to speak to the miners about unionizing the mine workers so that they would receive fair compensation for their labors. A number of representatives from each mine had been elected by their fellow workers and were authorized to discuss the needs of the mine workers. The union representatives would have information about the goals of the union, authority and membership. The meeting would be held in the Butte High School assembly hall at 7:30 that Thursday evening.

Marcus threw down the paper, stomped out of the office and went back to the living room. Everyone was still seated visiting when they heard Marcus coming down the hall talking himself.

"Margaret, those damn Easterners are going to have a meeting with our miners to try and organize them. No union is going to tell me how to run the mines, and no miner with Anaconda needs a union to be treated fairly. I've always given them good wages. It's because of Anaconda Copper that the other miners get better wages than is general.

"I'm going to that meeting in Butte. I have a few things to say to those rabble-rousers that stir up our good men."

"Well, Marcus, I suppose you must, but please calm down. You'll have an attack sometime if you keep letting yourself get so upset. You will pardon Marcus, my friends, won't you?"

Then Marcus looked around and saw all the people looking at him.

"Well, I'm sorry, but that really angers me. I do everything I can for the miners, and they have a good life. Hell, they never would have survived California, where I did my time in the mines. I'm sorry I blew off to you people."

Mrs. Lewis said, "Marcus, we're family and we certainly aren't offended by your remarks. We understand, and we know you care a great deal about your men. I really don't think your miners would join a union."

"Well, maybe not, but I'm still going to the meeting to

have a say. I'll leave you now and retire. Probably I'll go to Butte Wednesday."

The outburst ended the evening visit, and everyone decided to retire to their rooms.

Margaret said, "If you want to stay here you certainly may. There are books in the bookcases along the wall. I ordered some of the most recent novels to be sent here and of course, there are the classics. The days were full of activities at the mansion, so everyone said they were ready to retire.

Margaret stopped at Marcus' room to see if he had calmed himself.

"Marcus, dear," Margaret called as she rapped at his door.

"Come in, Margaret. Sorry I was such a bore. I did very well until I read that article in the paper. I'm still burning."

"Marcus, I'm sure nothing will come of it. Your men are devoted to you. Don't get so upset — it isn't good for you. You've heard of bursting a blood vessel — well, that is what's going to happen to you if you don't quit exploding."

"All right, Margaret, I'll try. Don't forget we're going to the sawmill tomorrow. I want you to see it and Mark is interested, too. Could we leave around 9:00? The next day I'm going to Butte to find out what's going on," said Marcus.

"All right, dear — yes — 9:00 is fine. Mark will be ready. Are you all right now, Marcus? If so I'm going to retire. I just want to be sure you're settled," said Margaret.

"I'm fine. Good night, Margaret, me darlin'." Marcus put his arm on Margaret's shoulder and gave her a kiss on the cheek.

"Good night, Marcus. I'll be ready in the morning."

At 9 a.m. the carriage pulled up into the north portico, and

Marcus, Margaret and Mark came down the steps and settled in the carriage. People in town would know that this team of horses was from the Daly mansion. The team of horses were matched blood bays. Horses and wagons or carriages could be easily identified, for each farm had different marked horses than the other so that Mr. Daly would know immediately who was in town.

Mark sat up with the driver and said, "Papa, I can see all over when I'm up this high. I love it."

Margaret and Marcus sat behind in comfort and grandeur. The carriage was "The Victorian" and was new to the Mansion carriages.

"Marcus, this is really the most comfortable carriage, even on these bumpy roads. I like it very much."

"I'm glad, Margaret. It's new this year, but I ordered it some time ago. I'm pleased with it, too. Certainly, Mark likes riding high," said Marcus.

As they went along to the sawmill Marcus told Mark and Margaret of the future plans. The street they were on now would have a railroad spur to bring the lumber to the main tracks. One street would have tracks going one way and the next street would have tracks going the other. The sawmill came into view. It was a large operation and covered a number of acres. The river was to the west of the mill. The river bed sloped down from the mill, with the mill set on ground high enough not to flood with the annual spring runoff from the snowpack in the Bitterroots.

Margaret and Mark watched as the logs came in. The bark was stripped from the logs, cut into board lengths and loaded on to the railroad cars. Mark said, "Papa, the men work hard and fast here. There sure is a lot of noise."

"Yes, Mark, the work is hard."

"How do they get the logs here, Papa?"

"There are three ways to get the logs to the water. The spring runoff fills the creeks with high, rushing water. A slough can be built to direct the logs into the creek, ending in the river. Sometimes they use wagons with men and horses braking the wagons down the mountain until they reach the river. The third way is fast and hard on the horses and drivers: a smaller wagon is loaded with logs, then the driver and horses come pell-mell down the mountain to the river. This last method is used when it is a short distance to the river. The logs float down the river to a holding basin and stay until the mill is ready to use them."

"Do they cut the trees for logs up in the mountains? Do the people go from Hamilton each day to work in the forest, Papa?"

"I can see that you are an inquisitive one, Mark. That's good. You should learn all you can about business and the people that work for you. There are logging camps with living quarters, cookhouses and shelters for the work horses. There is one camp up on Roaring Lion west and south of here. The camps open as soon as the snowpack on the logging roads melts enough to make it passable. The camps are closed when the snow returns for the winter. There is a six-month period that the forest is harvested. It's a big operation, Mark. The camps will have forty to sixty men and thirty to sixty horses.

"Anaconda Company contracts the timber harvest to logging jobbers. Anaconda owns the camps. This summer, railroad spurs will be built to a central area and the logs loaded on railroad cars. The loaded cars with the logs will go directly to Butte and Anaconda by rail to be used to shore up the mines in Butte and for fuel at the smelter in Anaconda," said Marcus.

"I'd like to go to the camp," said Mark.

"Perhaps you can later in the summer." said Marcus

Margaret had been listening with interest. She had gone with her father on lumber sales and had heard tales of the lumber camps. Margaret remembered well one sale of her father's, when she first met Marcus Daly.

"Well, Margaret, I'm going to go with Jim Hamilton to the office in town. You and Mark can take the carriage into town and then home. Jim will see me home. I would like Jim to stay for dinner. I hope he has time to come."

"That's fine, Marcus. I'm sure Jim will come if you don't overload him with another project," said Margaret.

Marcus laughed. "I know I'm a pusher, but there is so much to do and Jim feels the same way. We feed on each other's enthusiasm and ideas."

"Well, Mark, can you take care of your mother and see that you both get home safely? Take your mother through Hamilton. There are some new buildings. The company store is open and well stocked. You might take your sisters some hard candy."

"All right, Papa. Can we stop at the store, Mama?"

Margaret was glad that the store was open -- she wanted to see the different items offered besides groceries. "Yes, Mark, and then we must hurry home. We will be too late for luncheon and too early for tea. Stena put a box lunch in our carriage and we will stop at the pond on the way home and have a picnic."

Mark smiled and felt very proud that he was to take care of his mother. "Papa, thank you for taking me with you. I liked seeing the mill. I'll see that we get home safely. Bye, Papa."

Marcus waved to Margaret and Mark who were looking back at him from the carriage moving quickly towards Main Street and the town of Hamilton, one mile away.

The next day, Marcus boarded the train for Butte, intending to go to the Union organizational meeting. Marcus had talked to the manager of the Anaconda mines, who told Marcus not to go to the meeting. Management would be barred. Only the mine workers and mine support workers could attend. "That burns me up. Isn't this a free country? I'm going to that meeting. I want to know what rot they are telling my men. I'll take care of our workers. We don't need any damn Eastern ne'er-do-wells out here telling us how to run our business."

"Mr. Daly, I think you are asking for trouble, but I'll go with you to the meeting."

"No, you won't, I'll just walk right in. If I'm by myself, they won't know me."

As Marcus approached the school, he saw a large crowd gathered. He joined the group staying behind and trying not to be recognized. So far, so good, thought Marcus. The line started going into the building. When Marcus came to the entrance, two men stepped up to him and said, "Mr. Daly, you are not allowed at this meeting."

"Why am I not allowed? I've been a miner. I know what it is to work. I want to hear what you are telling the men."

"Mr. Daly, we will remove you forcibly if need be."

"You will not. I don't want to start a riot, but if you lay one hand on me, my men will come out swinging."

Just then two of Marcus's miners came to him and said, "Mr. Daly, come with us. We don't want you to be hurt or insulted.

We have something to tell you."

Marcus turned to them and walked away from the door with them.

"Mr. Daly, don't you worry none. Our miners will not join the union. We don't need or want it. We will tell you what goes on. You'll know everything just as if you were there."

Marcus decided that perhaps they were the wiser heads to prevail and agreed to leave the meeting site. But never had his feelings been more hurt nor had he felt more helpless.

Marcus went to the Butte Hotel to wait for the outcome of the meeting. Just as Marcus was about to give up and go to bed, the men came and reported to him that the Mine Workers Union had been embraced by most of the mine workers.

"Mr. Daly, the workers at Anaconda mine did not vote to join the union and will not be unionized. The other mine workers joined and the United Mine Workers Union was established with Butte being #1 Union. The unionizers said there will be hundreds of groups joining the United Mine Workers Union in the future.

"Mr. Daly, you can count on your miners. As long as you are running Anaconda Copper there will be no union at the Anaconda mine."

"Thanks, lads. You can be sure anything the union can do we can do better and quicker. Keep your eyes and ears open for union news. We can give benefits before the union has time to act. I appreciate your support," said Marcus. "The unions will be hard pressed to keep up with Anaconda Copper Company salaries and benefits."

Marcus left Butte the next morning on the train to Anaconda. He talked to his old friend Murray about the union meeting in Butte and the resulting frustration.

"Sorry you had to go it alone, Marcus. Why didn't you let me know?" said Murray.

"I thought I could pull it off by myself. At any rate, our men didn't join, but I'm telling you, the Union will be the ruination of the industry. I hope I don't live to see it," said Marcus. Let's do a little business here, Murray. I want to leave tomorrow to go back to Riverside. Are you coming over some weekend? We're having some horse races early August. The races start here in Anaconda and Butte in late August. Just come over anytime, Murray."

"Okay, Marcus, to work. All is going well here, I think. Let's see what you think."

"As to the visit -- not this season. I appreciate you're wanting me to come, and I'm anxious to see the operation at Riverside now that it's complete; but the smelter needs all my attention now. I'll go to the races here. I hear you got some good thoroughbreds on your trip."

Marcus and Murray continued talking about horses, smelters and unions as they walked from the office over to the smelter. After the smelter tour Marcus went to the Montana Hotel. "Murray, come have dinner with me tonight. We won't talk shop — we'll talk about old times," said Marcus.

"All right, Marcus, let's not get tied up in one of our old-time bar scenes," said Murray.

"No, Murray, I left those days when I married Margaret. I still like to have an evening with the boys, but too much drink is hard on me. With each drink I thought I got smarter and finally I was so smart someone would take a poke at me and we were off," said Marcus.

Murray laughed. "I remember, Marcus, I was there!"

"Seven o'clock, then. I'll meet you in the bar at the hotel."

The evening went quickly. Marcus and Murray through the years seldom confided in one another. They were the closest of friends and understood each other without speaking.

"Gad, Murray, can you believe all that has happened since we first met in California? There's still so much to do, but that makes the world more interesting, eh, Murray?"

"I suppose. But right now I would like to slow down a little."

"Do it then. You're your own boss and you have good men. Take time for yourself. You have my blessings, and you deserve time off."

"I might. I'll let you know when. Tonight has been good, Marcus. It doesn't seem as if twenty-eight years have passed since we were young lads together in a mine in California."

"I'll see you next time around, Murray. Take care of yourself." Marcus went through the lobby to the elevator and turned to see Murray leave the hotel.

I've had many acquaintances, but few close friends, Marcus thought to himself. Murray and Haggin are the only ones, but two better friends no one ever had.

Marcus smiled, opened the door to his room, settled into the bed, and with a sigh, was asleep immediately.

The next morning, before dawn, Marcus awoke. It was 5:30. There was a built-in alarm system in Marcus and once awake he couldn't go back to sleep. Nightmares came in waves if he went back to sleep, leaving him more exhausted than if he had gotten up too early. The train to Butte was on time. The stop in Garrison to switch to the Northern Pacific line was, as usual, a time-consuming process, but Hamilton and Riverside would be at the end of the train line, so Marcus would be home in one day. Others were waiting at Garrison and a stranger next to Marcus spoke to him.

"Mr. Daly, pardon my interruption, but I'm a reporter with the Helena newspaper and I wondered if you would give me an short interview, please. My name is James Frost."

"Well James, I guess it won't hurt. What do you want to know?"

"Mr. Daly, I understand you purchased some trotters and racing horses in Europe."

"Yes, I did, twenty-five of them just arrived from Europe and are in quarantine until we are sure they are sound. Then we will start training them for the races in Hamilton, Butte and Anaconda. Our first races will be in August."

"Mr. Daly, do you like New York City?"

"Well, lad, my family spends time there, so I like New York because I can be with them. I like the New York Democratic party and have become very interested in national politics. You can get more done politically in the large population of New York. Numbers mean power, Mr. Frost."

"Mr. Daly, where do you spend the most time?"

"Well, I think I spend the most time here at Garrison, waiting on the damn trains!"

Mr. Frost threw back his head and laughed and said, "Mr. Daly, I guess that ends our interview, because the train's switching cars and we'll be on our way. Thank you, Mr. Daly."

"T'was nothing boy — nice to meet you," said Marcus.

The train pulled into Riverside before dark and the carriage with Eddie was there for Marcus.

"Evening, Mr. Daly."

"Evening, Eddie. Is all well at the house?"

"Yes, sir," Eddie turned the carriage up the lane toward the mansion, arriving at the north portico. Marcus entered the mansion and went directly to his room to bathe and ready himself for dinner. Refreshed from his bath, Marcus felt something inside him that gave him a zest for life. He had that feeling often, an excitement as if something good was going to happen. Generally nothing out of the ordinary did happen, but it gave him a lift.

Marcus stopped at the children's dining room and disrupted their dinner, which they loved. Nobody else ever did that. Then it was on to the living room where the family and guests were having cocktails before dinner. Marcus greeted Margaret first and then turned to greet the Lewises, the Evanses, Mattie, and Dr. and Mrs. Murray from Butte. Dr. Murray had a medical clinic in Butte and was the Daly family personal physician and friend. The summer would not be complete without Dr. and Mrs. Murray and their two children at Riverside. A suite in the mansion was reserved for the Murray family. The days passed quickly and soon it was August. The race tracks were finished in Hamilton, Butte and Anaconda. As promised, the horse races would be open to the public and take place on the Riverside racetrack. The first day would be thoroughbred horse racing. The second day the trotters pulling sulky and driver would be held.

"Such excitement. Imagine having horse races right here in the Bitterroot," said Dr. Murray.

Betting, companionship and horsemanship made the races the most exciting outing in the Bitterroot. The entire valley came and enjoyed.

There were signs telling of the races in Anaconda. The dates of the races were posted. Under the racing information the Montana Hotel was listed as having accommodations for families, and the Florence Hotel advertised as a single man's hotel.

When the last two weeks in August came the family boarded the private cars and went to Anaconda to see the horse races. Mr. and Mrs. Evans would not return to Riverside until the next year. They would open their Anaconda house to the Daly family. The overflow of guests would stay at the Montana.

It was a festive time. Marcus gave the miners the day off and a day's pay so that they might attend the races. Marcus went to the first day of the races at Anaconda. He went to the paddock to find Sam Lucas. "Sam," Marcus called, "Sam. Have you got a minute for me?"

"Yes, sir, Mr. Daly, what can I do for you?"

"Sam, how are the horses doing? Do you have the times on horse races in New York? Do you think that by next spring we could enter some of our horses in the eastern race circuit?"

"Why, I think so, Mr. Daly. I know the times from last year and our horses measure up, but I don't know how they would take the train ride all that distance."

"Hell, Sam, half of them came over on a ship from England. I don't think the train for three days will be too hard on them. We'll take them early enough to acclimate in the east. You know I have the farm in Maryland. We'll take the horses and the timothy hay that we grow here in the Bitterroots. The hay is more nutritious and I want to keep their feed the same. The veterinarian will accompany the horses on the train and stay with them. The trainers and jockeys and you could make those horses think they are right at home."

"That's an ambitious project, Mr. Daly, but, by gad, I think we can do it. If you keep the support system for the horses the same, they should be just fine. That is really an exciting plan, and if I'm any judge of horseflesh, I put your stable in the same category as Belmont's. I'm really pleased with your thoroughbred selections."

"Thank you, Sam. I consider your opinion that of an expert. I'm going to the stands now. I hope our horses show off today for this crowd."

"I think they will. The horses were fine getting on and off the train and rode well in the railroad cars. That is hard for animals. They've had a good rest and will be ready to go on the new race track. I see the tracks you've had built are the same as Belmont, in the shape, the smooth hard surface and the one-and-a-half-mile length. Our horses should run well. They'll think they are on their own track at Riverside. See you after the races, Mr. Daly."

Marcus went to the stands where all the family was gathered. On the way Marcus spoke to some of the men from the mine and smelter.

"Thanks for the day off with pay, Mr. Daly. We're going to enjoy this day at the races. Which horses should we bet on, Mr. Daly?"

"I never tell anyone how to bet on the horses. I think it is bad luck for my horse, for you and for me. All the Daly horses are well trained and well bred. Just take your pick."

Marcus entered the stand and saw Margaret. He settled himself in the seat beside her. Marcus glanced around and saw W.

A Clark. Marcus' and Clark's eyes met for a moment, and a flicker of the eye showed recognition, but they both chose to pretend they had not seen each other.

"Did you see who was here, Margaret? Old William Clark. I didn't think he would come to the races, since I'm involved," said Marcus.

"Marcus, this is a social event and everyone is here to see the races, not to know who owns what and who likes or dislikes whom," said Margaret.

"Whoa, Margaret, I didn't mean to get you riled. I was just surprised. I'm sorry, Margaret," said Marcus.

"That's all right, Marcus, let's just enjoy the day. It's a bit warm, but the shade in this grandstand makes it very comfortable."

The races were a success. The next day they would see the sulky races. As advertised in the newspapers, the sulky races were just as successful. The day after found the Daly family in their place in the stands.

"Margaret, did you know that sulky racing is the most efficient science? The horse trots in perfect rhythm, and the sulky and driver aid the animal's performance, all in perfect unison," said Marcus.

"I hadn't thought of that, Marcus. I do know they are beautiful and exciting to watch," said Margaret.

The young people each selected a favorite. The girls liked the Irish horse in the Daly colors. Mark liked the horse and driver from Maryland in the green and gold Daly racing colors. They couldn't remember the horses' names, so Molly called out, "Come on, Ireland" and Mark yelled "Come on, Maryland."

Marcus laughed. "Margaret, the children have a code of

their own."

A grand time was had by all -- the wealthy, the workers with their special holiday, and the young people roaming free and mingling unsupervised with each other.

Madge and Molly overheard a conversation in the stands. "You can say what you want about Marcus Daly, but these races are a great asset to the communities. We are watching the finest horses racing right here in Butte. It wasn't many years ago the Indian wars were being fought. Now we have a civilized sport right here for our pleasure," said the first gentlemen.

"It is amazing what change the last ten years have wrought."

"Understand these horses of Dalys are going to New York to race at the Belmont track," said the second gentleman.

"I didn't know that."

"Well, it may be a rumor. Sounds like a mighty ambitious project to me."

"I think they'll do well. Certainly, Daly has spared no expense on horses, trainers, and a healthful environment for the horses. I guess his Bitterroot horse farm has everything old Belmont had on his horse farm in New York."

Molly and Madge blushed with pleasure. "We must tell Papa and Mama when we get back to Grandmother's. Papa will be pleased," said Madge.

The summer ended abruptly with a drop in the temperature on September 4th and the departure of Mrs. Daly and the children for New York City.

Chapter Nine

The next nine years were the best and the worst for Marcus Daly.

The best was a return each year for the summer at Riverside with family and friends. The thoroughbred horse racing and sulky racing were great events in the Bitterroot Valley. The horses were recognized as winners on the eastern racing circuit, and the horse farm was a productive and profitable venture. The mines in Butte were pouring millions into Anaconda Copper and the partners, including Marcus, were among the wealthiest people in the world. The smelter paid for itself and the town of Anaconda in a few short years.

The dark aspect of this time involved A. W. Clark and his political feud with Marcus Daly. (The copper owners of mines and the "apex theory" were taking hours and hours of litigation and testimony, as well as millions of dollars.) Marcus was quoted as saying "without the Riverside farms to come to and relax, I would not be able to carry on." Margaret continued to take the children to Paris in January to further their schooling, returning to New York each spring to meet the demands of their social life. When Margaret returned to New York she engaged a patron to introduce Madge and Molly into society. She wished she had done it sooner so that there would have been more time before Madge was eighteen. When Margaret told Madge and Molly about their patron, Madge said, "Mother, I don't want to do that. I don't know any of these people."

"My dear, I'm doing this so you will meet the right people. You know there are people who are fortune hunters, and I don't want you to become prey for them. Your sponsor will screen people, and she has access to the finest homes and parties. And, Molly, she will have two years to introduce you before your coming out at eighteen."

"Who is our patron?" asked Molly.

"The one I've engaged is from an old New York family. She is accepted in society and whomever she introduces also will be accepted. Her family has lost their fortune, but they still have their good name. She is a lovely woman and she will take good care of you."

"Oh Mama, I'll be so embarrassed!" said Molly.

"No, you won't. It's done all the time. We're fortunate that this patron would accept us. She has many noveau rich seeking her help," said Margaret.

"What is her name?" asked Madge.

"Her name is Mrs. Abbot Cornelius. You will meet her next week at tea. Relax -- you have all the necessary training to be ladies, and Mrs. Cornelius will do the rest," said Margaret. Since Madge and Molly were dutiful daughters they accepted their mother's wishes without any rebellion. Marcus was getting ready to leave the Bitterroot. He had done his work here and it had been a satisfying time for him. He went to the mansion to pack for the trip. First he would deal with his office in the mansion. He would answer his latest packet of mail before he left. There was a letter from Ben Haggin.

"Dear Marcus,

Your horse farm should be ready for the thoroughbreds and training. My best mare had a colt that I think you might be interested in buying. Both the mare and stallion have excellent blood lines and this colt loves to run. Since he was born and could stand up he has led the other colts in all of their runs together. He is a yearling now. Let me know if you are interested. I have too many yearlings and I need some winnings on the track.

I'll be in Butte last of this month and I will see you there. I can bring the colt along if you are interested.

Your friend,

Ben Ali Haggin"

Marcus immediately sat down and wrote to Ben Haggin. He had a feeling about this colt. If Ben said the yearling was special, Marcus believed him. Ben had an impressive stable of race horses.

"Dear Ben,

Send the yearling. You didn't quote a price, but I assume you will know by the time the colt is delivered. Bring him when you come. I'll meet you in Missoula and transfer the colt to the Valley train and for the trip home to Riverside.

If you want to come to Riverside, I've decided to stay here longer and I will keep the mansion open until I leave for New York in October.

Thanks for thinking of me, Ben.

Your faithful friend,

Marcus Daly"

Ben Haggin and the yearling arrived in Missoula on September 23, 1890. Marcus and Ben greeted each other with warm handshakes. "Ben, good to see you — you look great. California agrees with you. I'm anxious to see the yearling."

"Lets go, Marcus. He's right down here and they are going to unload him shortly."

As they came up beside the car with Haggin's yearling

aboard, the ramp came down from the freight car and a stable boy walked down it, leading the colt. Marcus watched as the yearling cautiously stepped on the ramp and calmly walked down; but, when no one was expecting temperament, the colt kicked at the stable boy.

"A spirited one, Ben. He'll have a good home at Riverside and we'll make him the best race horse that will use his potential. I like his looks. His manner is spirited, I'll say. I'll name him Montana."

"Ben, are you coming on to Riverside?" I would like for you to see the horse farm."

"Yes, Marcus. I want to see the whole horse operation and the other farms, too."

Marcus and Ben spent three days at the mansion and horse farm. Ben said, "Marcus, you've a fine horse farm and probably the only one with its own veterinary hospital and two resident veterinarians. You should do well with this stable."

"We'll see Ben. I'm anxious to go to the Belmont races. Say, Ben, you haven't seen the race tracks at Anaconda and Butte. They can seat 10,000 at Anaconda and 20,000 at Butte. Our purses will be equal to the best-known tracks. We'll have the best horse blood in America with the race enthusiasts coming from all over America and Canada," said Marcus.

"Marcus, you mean the tracks are finished?" asked Ben.

"Yes, Ben. We had races there in August. Next year we'll have races thirty days in Anaconda and a sixty-day meet in Butte."

"Unbelievable, Marcus. That is unbelievable," said Ben.

When Ben left on the train, Marcus stayed at Riverside another two weeks. When satisfied everything was in order,

Marcus had the mansion closed for the winter and traveled by regular train to Butte. Marcus only used the private cars for family and for trainers and veterinarians when the horses were transported to the farm in Maryland.

Nothing was too good for the horses. The horse barn was heated and the stalls lined. As Marcus used to say, "If I kept the horses in the house and I lived in the barn, I would have the better deal."

Marcus met Ben Haggin again in Butte. They discussed Anaconda Company business and toured the facilities at Anaconda and Butte. Marcus told Ben about his episode with the union organizers. "Well, at least our miners aren't unionized, Marcus and I know that is because of you," said Ben. "I'm holding off Standard Oil and keeping Tevis and Hearst from selling. As long as you are making the mine and smelter pay so well, I want you to have control. I'm afraid Standard Oil would interfere with your decisions. Or, at least, they would delay, study and talk before any action. Sometimes that is too late. I'll keep you posted on my success with our partners. Hearst is in newspapers because of his son, and Tevis says he is getting old and wants to gather his resources. He's no older than I am. What will he do with more money? Its ridiculous," said Ben.

"Well, I appreciate your holding out, Ben. I know there will be hell to pay when Standard Oil gets control. The men will unionize for sure, and we have this young man Heinze that is well liked by the miners and women and disliked by the mine owners. He's a clever one, and he'll agitate everyone one way or the other. Right now he is in litigation with Boston and Montana mining company using the apex theory to claim ore from the Boston mines. I'm watching that one. Heinze knows all the angles and uses them. He has won a judge over to his side by being the "poor little mine owner" being attacked by the big syndicates and corporations. It's hog-wash. He plain mines next to the corporate property, takes the ore as he pleases and rushes it to the smelter," said Marcus.

"We'll need to watch these events closely, Marcus."

They decided to meet again the following summer, 1891, in Butte. "Let's make it August when the horses are racing," said Marcus.

"It's done," said Ben.

Marcus left for New York the next week. Ben left for California. Margaret and family waited for Marcus at home for he was to arrive from the west in time for his birthday, November 12.

Marcus had settled in at the Nederland Hotel after his summer at Riverside. He made visits daily to his family at 725 Fifth Avenue. That night was his birthday, and he busily dressed for dinner and the party with Margaret and children. It was not very often that Marcus was available for his birthday because his travels took him far and wide. One thing Marcus vowed was that Christmas would always be with the family. He never missed Christmas, the most favored holiday in the Daly family.

Marcus walked from the Nederland hotel to the apartment, entered the building and rode the elevator to the second floor. Cutler opened the door, greeted Marcus and escorted him to the living area where the family was gathered.

"Papa, happy birthday," said Madge and Molly in unison.

Margaret rose from her chair and came across the room to meet Marcus. "Good evening, Marcus dear, you look wonderful.

I'm so pleased as that shows you have been taking care of yourself since we parted."

"You're right, my dear. I've been working, eating and sleeping the prescribed amount," said Marcus.

Harriot and Mark were standing just behind Margaret, and Marcus leaned down and hugged Harriot and patted Mark on the shoulder. Harriot said, "Papa, we've told everyone about Riverside, and I told them about my horse, Shamrock. They think Montana is somewhere outside the United States."

Mark said, "I set them straight about Montana. Now I have a U.S. map I carry to show them Montana."

"Well," Marcus said, as he looked at Margaret. "That's New York!"

Cutler came to the doorway and announced dinner.

Harriot started singing, "Happy birthday to you, happy birthday to you, happy birthday, dear Papa, happy birthday to you." They proceeded to the dining room to the tune of Happy Birthday. Marcus loved it.

The dinner was substantial, but not as much as usual, for the birthday cake and ice cream were special. The cake came at the end of the meal with forty-eight candles ablaze.

Mark said, "Papa, will you be able to blow all those candles out?"

"We'll see, Mark. I think I can." Marcus blew them all out with one breath and all clapped in glee.

"What a nice party you've had for me," said Marcus.

"Wait, Papa, we have a present for you. This is the latest entertainment." said Mark, "It's a zoetrope."

"A what?" said Marcus. He looked at a little black barrel-shaped object with an opening at the top and slits on the side, set on a wooden base. The drum rotated on the base.

"How does it work, Mark?" Mark inserted a strip of animated cartoon animals and rotated the drum, creating a illusion of motion. There were a number of strips with the gift, creating a different series of stories.

"Now you won't be lonely, Papa. You can look at pictures in your zoetrope." said Harriot. Her tongue tripped over "zoetrope," but she finally got it said.

"Thank you. I've never seem a zoetrope, and I think it's a most thoughtful and interesting gift. Let's do some pictures," said Marcus.

After all had taken their turn looking, Margaret took a box from her shirt jacket. "Here, Marcus, I have something for you that I admired and I want you to have from me."

Marcus opened the box and saw a diamond stick pin.

"Margaret, that is a fair-sized diamond! You must have saved your household money. No, me darlin', I don't mean that. I like it very much and will be proud to wear it. Thank you darlin'."

Marcus left the apartment about 10:30 and walked to the hotel. He had appointments in the morning and later he would meet with Margaret and hear more about the children and herself.

Marcus and Margaret sat in the living room at 3:00 the following day and Margaret began to tell Marcus their news. "Marcus, I have found a friend who wants to sponsor Madge for a coming-out party with the other eighteen-year-old girls of prominent families."

"Did she do it out of the kindness of her heart, Margaret, or did you contract to pay her?"

"Now, Marcus, don't get upset. You hire or contract people to construct buildings, plants, race courses and the like for you, and you pay them. I'm contracting (as you say) this woman to help construct our daughters future. The people she will be meeting will be as wealthy as she, and hopefully I won't have to worry about fortune hunters. I would be devastated if my daughter was chosen for her wealth instead of for herself."

"Well, Margaret, this woman may offer some protection, but I wouldn't count on that. If you hired her to introduce Madge to eligible people, that is what she does. I think I have enough business acquaintances that have eligible sons and daughters that the girls could meet."

"No, Marcus, you must be a debutante in New York to be in the right social group."

"All right, Margaret, have it your way. I hope that Madge is happy with the decision. How about Molly?"

"Well, Mrs. Cornelius will have two years before Molly is of age, and by that time I'm sure Molly will be part of a group of friends and able to enjoy the parties."

"What can I say? I'm dumbfounded. It's the craziest thing I have ever heard of," said Marcus.

"Now, Marcus, believe me, I know what is best. I agree that it does sound insane, but it has to be done for our children's sake. I can't do it alone, Marcus," said Margaret.

"Fine, Margaret, fine. Is there anything else concerning the children that needs my attention?" said Marcus.

"Not really, Marcus. Harriot and Mark are still young and they will make their friends through the years by attending the best schools here in New York," said Margaret. "I know you are upset,

Marcus, but I'm making an investment for our daughters' future. You understand investments, don't you, Marcus?"

The statement took Marcus by surprise. He had not thought of Mrs. Cornelius as an investment, but he did realize that Margaret was only interested in giving Madge the best opportunity for happiness.

Marcus still didn't like the idea of a patron -- a paid one at that -- but he would not presume to tell Margaret what was best for the children. Margaret was like a tigress and she would fight the world for her children.

As Marcus left for the night Margaret sensed his disapproval of her plans. As she kissed him on the cheek Marcus said, "Margaret, you are probably right about your plans. Don't mind me. I'll see you tomorrow."

The weeks went by, and Christmas came with all its decorations and festivities. There were glorious days in the park with Marcus, Margaret and the children. The younger ones walked briskly in front of Marcus and Margaret, while Madge and Molly strolled behind with solemn sophistication, ignoring the foursome ahead of them.

Mark and Harriot were throwing snowballs at each other, avoiding the urge to pelt Madge and Molly, who thoroughly disapproved of the youngsters' antics.

The year 1891 brought a new fashion trend. The pattern books from Paris and the *Harpers Bazaar* magazine arrived with the latest fashions. Cutler collected the mail and brought it to Margaret. "Thank you, Cutler. Would you please tell Miss Madge and Miss Molly to join me here in my sitting room when they come in from their classes?" asked Margaret.

"Of course, madam," said Cutler.

As promised, Cutler met the girls at the front door and relayed Mrs. Daly's message to them.

Molly said, "Goodness, have we done something wrong?"

"I hope not, I can't think of anything. We've not had any bad reports at school. Let's hurry along," said Molly.

"Dear girls, come in, I just wanted you to see the new dress and coat patterns from Paris and the pictures in *Harpers Bazaar*. It is a nice change. The bustle is smaller but so is the waist. They call it the hour-glass figure. Look, the corset has laces so that the waist can be cinched in to a smaller size," said Margaret.

"The figure looks very nice, but I wonder if the small waist will be comfortable."

"Let's select some patterns and I will call our dressmaker to come along as soon as she can," said Margaret.

"Mama, look at this one. Wouldn't this be nice for Harriot?" said Madge. "It's pink and white plaid with a black faille frock coat (just like the adults). The sash looks as if it's faille too, Mama. The hat is straw and has a double pink satin bow on top. Actually, the hat looks like a pancake with a back turned up," said Madge.

"That hat doesn't sound attractive," said Margaret.

"But it is, Mama. Here, look at the picture."

"Yes it is, Madge. You're right, the hat is pretty. Actually, this issue of *Harpers* has beautiful clothes. I would like any one of them," said Margaret.

"For myself, I like the one next to Harriot's dress." said Margaret. "It has a black silk jacket with a double collar and a large lace collar under the black silk collar. The dress is black and white striped satin. I like the high neckline, tied in the back of the neck with bows. That's a new idea."

"What a relief. The hats are smaller. We were always bumping hats in the carriage, weren't we? I'd like the hat pictured with the outfit. It's a small black hat with silk ruching around the brim. It is worn forward on the head and has lovely flowers on the top. I will have that outfit."

"Molly, Madge, which one do each of you want?" asked Margaret.

"Listen to this one," said Molly, and she began to read from the picture. "A graceful gown of India silk beige with black pinstripes background and strewn with pink and yellow blossoms. The full bodice is fastened on the right side under a jabot of black lisse. Black satin ribbon edges the bodice and drops in loops in the front. The straw hat is trimmed with pink ribbons and pink and cream colored blossoms. Mama, could this one be my spring dress for nice occasions?"

"That sounds lovely, Molly. Bring it over for me to see. Yes, I think that would be perfect for you, Molly," said Margaret.

"This one is pretty, said Madge. "I'll read to you what it says. 'This is a bordered gown with a border half the depth of the skirt. It is of pale blue wool with ecru bands. It is finished at the foot with black bengaline stripes. Wide point de Gines lace forms cuffs on the sleeves and a cravat at the neck. Pale blue bengaline ribbon is used for the belt, sash and bows. The yellow straw hat is trimmed with lace and fine flowers.' Here's the picture, Mama."

"Lovely, Madge. You selected wisely. It will be beautiful on you."

"What about Mark, Mother?" said Molly.

"Well, dear, the boys' suits are standard, but I will have a lighter-weight wool tailored for him," said Margaret.

"We certainly will be grand this spring in our new clothes. Thank you, Mama."

"Oh, Madge, you'll need more clothes than this for your introduction into society in June. You'll need a wardrobe of party dresses, tea gowns, and sporting clothes. Take a day or two to select some choices from *Harpers Bazaar*," said Margaret.

"I'll do that this weekend, Mama. We must get our orders in, I know," said Madge.

"Thank you, dear," said Margaret, "you might look at the Worth gowns and cloaks, Madge. They are the finest and I will send an order to Paris."

Mrs. Cornelius was busy arranging for Madge to attend parties and become acquainted with other young women that were to be introduced in the spring at the New York Debutante Ball.

In February 1891, Margaret and the children went to Paris and took up residence until the spring parties in New York commenced prior to the Debutante Ball. Margaret and Madge took a great deal of time selecting Madge's wardrobe and particularly the white ball gown for the presentation at the ball. Madge was partial to the Worth gown pictured in the pattern book, and Margaret concurred. The gown was of white damask trimmed with embroidered lace and pearls. White tulle and lace were festooned around the neckline. Pearls and crystals were festooned diagonally across the front of the dress. The bottom of the skirt had a deep flounce of embroidered lace. Under the lace flounce was a ruche of tulle. With this gown Madge would wear a ornamental twist of pearls in her hair and carry a white ostrich-and-ivory fan.

Molly received the benefit of Madge's debutante wardrobe, for their mother couldn't bear to have Madge receive the lavish clothing and Molly receive nothing.

When the boxes, packages and trunks of clothing arrived at the Paris apartment, Madge, Molly and Margaret started the process of checking on the orders. Madge said, "Mama, the clothes are even more beautiful than I remember. Look at this by-the-sea outfit with its white trimmed sailor collar on the navy blue jacket and dress. I hope Mrs. Cornelius finds a gentlemen that I can enjoy when I wear all these different costumes. I will die if she brings me one of those impoverished gentlemen from Europe that have some sort of title. That is fashionable now. I can't bear it."

"Now, Madge, keep an open mind. Mrs. Cornelius has your happiness uppermost in her thoughts," said Margaret.

"You don't need to marry the person that is your escort. You can just be friends and enjoy each other's company. Don't worry so."

Molly had not said a word but took in all of the conversation between her mother and sister. She hoped by the time it was her turn to be presented to society, she would have a young male friend to escort her.

Marcus and Margaret planned to meet in New York the first part of May. Madge would be starting the round of parties and private balls that continued until the 1891 Debutante ball in New York in December. Mrs. Cornelius took charge of the social calendar and requirements for Margaret. Margaret and Madge just needed to be ready at the appointed time.

In April, Marcus went to Belle Meade, Tennessee, for the thoroughbred sale. He purchased Tammany for $2,500. Tammany's sire was the steeplechaser Iroquois who had won the English Derby. Tammany was bred by General W. H. Jackson in Tennessee and foaled in 1889. Marcus thought Tammany was the best two-year-old he had ever seen.

Tammany was sent to the Riverside horse farm. Marcus had one of the trainers accompany Tammany and he was to set up his training program. Marcus planned to enter Tammany in some races in the fall on the east coast. This would be Marcus's first horse to compete on the Eastern racing circuit. Marcus thought Tammany might be the one to prove his theory training horses in the Bitterroot resulted in more stamina from high altitude, the Bitterroot-grown timothy hay, and the excellent environment and training.

Marcus arrived at the Fifth Avenue apartment on May 12. He had just returned from the Bitterroot. Margaret was just finishing lunch in the dining room when Marcus entered.

"Well, Margaret, me dear, there you are just as I remembered. It's been a long spell. How are you?"

"Marcus, come here. I'm fine and you look very fit."

Marcus went over, put his arm around her shoulder and leaned down and kissed her.

"Margaret, let's try to arrange our lives so that we don't have three months or more pass before we are together," said Marcus. "How are the children? Your letters never tell me anything is wrong until it is all over."

"I know, Marcus. By the time my letter arrives the crisis would be over, so I don't bother."

Marcus laughed. "I guess you're right."

"Tell me about the horse, Tammany, Marcus,"

said Margaret.

"I think I got a good stallion, maybe a great one, for a reasonable price. He is gentle, loves to race, wants to please and is easy to handle."

"I wish you luck, Marcus. You've devoted a great deal of time collecting your stable. Knowing you, I know you want it to be a profitable undertaking."

"Me darlin', you know me well, but unlike mining, this enterprise of thoroughbred racing stable relaxes me and provides entertainment. I can breathe easier after I've been at Riverside with the horses."

"Then it is worth your time and money. If anyone needs a diversion from work, it's you," said Margaret.

"Marcus, Madge and I will be having a full schedule as soon as we return from Riverside this summer. She will be going to parties, and I will be here to oversee her activities. Molly is so busy observing whatever Madge does. Molly is going to be well prepared for her debut into society. Madge seems to enjoy sharing her knowledge with Molly."

"'Tis a blessing, me darlin', that our girls enjoy each other and have a sisterly love. It is a pleasure to be around them," said Marcus.

"Thank goodness, we have a number of years before Harriot's turn comes around. It really does require a great deal of the girls' time and mine, too." Margaret added, "Marcus, what are your plans for Riverside this summer? All the family is planning to come -- the Lewises, Dr. Murray and family, the Clarks and, of course, Mother, Father and Mattie."

"We'll have a grand summer, Margaret. There will be two

weeks of horse racing at each course in Hamilton, Butte and Anaconda."

"Mother said the theater in Anaconda is beautiful and there will be some activity there this summer. Marcus, I'm honored you named the theater "Margaret". I'm anxious to see it and to attend a performance."

"Since the railroad goes across the nation and the theaters are closed in the summer on the coasts, we'll get first-class plays and stars in Butte and Anaconda. I named it for you, Margaret, because I know women long for entertainment of the arts as much as we men love our sporting exhibitions. So it was for you and other wives that I did it."

"Well, thank you, my dear. I hope there are some men who will enjoy the theater as well," said Margaret.

"Sure they will. Margaret. If the ladies enjoy, we men can learn to enjoy, right?" Marcus grinned.

The summer was as promised. The Mansion was beautiful and guests arrived as planned. The routine that had been established through the years made for a relaxing time, offering plenty of action and entertainment for everyone. Activities offered for guests and the family were tennis, badminton, swimming, fishing, horseback riding, carriage rides through the estate, trips into Hamilton as desired and a playhouse for children only. The best part of Riverside was that it provided a place for a yearly family reunion. Mattie particularly appreciated her summers there.

In the summer of 1891, Mattie asked Margaret to come to her suite after lunch before they retired for their customary afternoon rest. Margaret rapped at Mattie's door and entered.

"Mattie, I hope nothing is the matter. I do so want you to

be happy here." said Margaret.

"Oh, Margaret, I just wanted to tell you, and you can tell Marcus, how very much I appreciate my accommodations. I know you have these two bedrooms on the main floor just for me and my companion. You even have the horses brought to the north portico for me where I can mount from the porch onto my horse. Everyone has a wonderful time here and loves your hospitality, but I know you have made special concessions in your plans so that my life will be easier and more enjoyable. Thank you, Margaret. No one is as kind as you and Marcus are."

"Well, that is nice of you to pay us such compliments, dear, but of course we want you to enjoy yourself here. Besides, every home should have a ground floor bedroom for various reasons. It is wonderful to have you come to Riverside, and I love having you here every year. Otherwise we would not have much time together," said Margaret.

"I just wanted you to know my appreciation for you and for the summer vacation at Riverside. The exertion needed to climb stairs and the resulting chest pains, has been eliminated here. Bless you, Margaret, for having all activities on the ground floor or outside in the gardens."

Margaret put her arm around Mattie's shoulders and hugged her to her side. "Mattie, it is our pleasure. You add so much to any company.

"By the way, we are taking two carriages into Hamilton. I know you will want to come. They have finished the dam on the Bitterroot River by the sawmill. They built a road across the top of the dam. It should give a good view of the river and holding basin for the logs. People can get back and forth to Hamilton via the dam."

"I will come. I haven't even seen the sawmill," said Mattie.

"I'll look for you at 3:00 then, Mattie. Get a good rest for our afternoon jaunt."

"Thank you, Margaret. I will rest," said Mattie.

The summer visitors left after the two weeks of horse racing in Hamilton. Marcus Daly's horses and other entries competed in Hamilton, Butte and Anaconda, and the races were well attended by horse-racing fans.

Marcus entered "Montana" in the Futurity. The odds on Montana were 40-1. Snapper Garrison, Daly's favorite jockey, rode Montana to a victory.

Tammany was entered in a number of races and had successful runs. Tammany was becoming Marcus's favorite horse.

Marcus went to the stables before leaving for New York and the fall racing season. The horses would be in their winter quarters as soon as the weather became severe. He wanted to see the two stables finished for the winter weather. The two stables had forty-eight stalls. Marcus called to Sam Lucas in the winter stables. "Sam, aren't these stables ready for use yet?"

"Yes they are, Marcus, but we won't use them until needed. I believe in keeping them outside as long as possible. I think they're healthier. Of course, after they come in for winter we still train and exercise them on the covered track. I have engaged twelve exercise boys."

"Sam, I want you to bring Montana and Tammany to race in New Jersey and New York this spring," said Marcus.

After the fall racing season in New York, Marcus knew he could no longer escape the social life of New York. Even some business receptions were social. He dreaded these but was always on his best behavior for Margaret and the children.

The President of Standard Oil, William Rockefeller and his wife invited Marcus and Margaret Daly to a reception. Because Marcus had dealt with most of the men on a business basis, he found the reception enjoyable. After a half hour he sought out Margaret. "Margaret, me darlin', this isn't bad. I'm enjoying the men. Are you getting along all right?"

"Yes, Marcus, isn't it lovely? The ladies I've met are very cordial and charming. I'm so glad you are happy, Marcus. I was worried," said Margaret.

"So was I, darlin'. I thought it would be a bad evening."

Margaret and Marcus enjoyed the autumn in New York. Madge was invited to numerous parties, and on the weekend before the Christmas Debutante Ball, Marcus was told by Mrs. Cornelius that he was to escort Madge. Madge wanted this, and Mrs. Cornelius thought it would be humiliating for Madge to have someone else escort her at the presentation ceremony. Mr. and Mrs. Cornelius would still be the sponsors along with Mr. and Mrs. Marcus Daly.

The night of the ball Marcus was dressed in the required formal dress of white tie and tails. Margaret wore a beige lace dress embroidered with peach and white flowers. She added three strands of her favorite pearls that were graduated in length, as well as a dog collar of pearls with a diamond brooch clasp. Madge had her beautiful white ball gown from Paris. The evening arrived, everyone was ready, and the ball was about to begin.

"Margaret, I'm nervous for Madge. Not to mention I'm a nervous wreck. I don't know why I let Mrs. Cornelius talk me into this," said Marcus.

"Marcus, you are doing it for Madge. Now please don't be nervous. You look very handsome and impressive. Most of these people respect you, Marcus, in business and in your successful racing stable, and I'm sure they will be pleased that we have a daughter being presented to society," said Margaret.

Marcus and Madge took their place and very soon heard, "Miss Margaret Augusta Daly of New York City, New York and Hamilton, Montana, escorted by her father, Mr. Marcus Daly, and sponsored by Mr. and Mrs. Abbott Cornelius of New York City, New York."

They walked across the ballroom arm in arm as the audience applauded and Marcus breathed a sigh of relief. "Well, Madge, darlin', we're almost across the floor and then you can relax and enjoy yourself," said Marcus.

"Papa, I'm so glad you could escort me. I know you don't like social parties, but it makes Mama happy and me. too. Thank you, Papa." said Madge.

Marcus and Madge acknowledged the introductions, and Madge joined the line of young women who were the debutantes of 1891. The young women would officially take their places in the world.

The ballroom was dazzling with chandeliers of the finest crystal and ablaze with the new electric lights. The decorations were in white and silver. The walls were festooned with white gauze sparkling with silver glitter and tied with huge bows. Garlands of flowers swagged below the gauze following the same pattern as the gauze. There were planters with gardenia bushes and trees with silver bows and white candles alight on the fir trees in the background around the hall. The chandeliers were swagged between with garlands of silver bows, white flowers and gauze with glitter. It was a winter wonderland. Each debutante was dressed in the finest and latest fashions. It was a glorious sight. One hundred young women were introduced into society, and the music from the orchestra and the dancing began.

Madge danced with her father around the perimeter of the dance floor and the next dance started the use of the dance cards. Each young woman's card had been filled before the evening's event. No one was to be a wallflower. It was an evening to remember. The eligible young men knew their duties and looked forward to the evening and the finale of the year of parties and meetings of young people. By the time the ball date came, almost everyone knew each other. This made for a gala evening.

Mrs. Cornelius was pleased, for Madge had made friends and was well liked. Madge had not found any young man to be more that an acquaintance, but she enjoyed the evening and she did have a full dance card. "Margaret, did I do fine for you and Madge?" asked Marcus. "It wasn't so bad, and Madge said I gave her the extra moral support she needed. Madge looks grand, Margaret. We can be proud of our firstborn."

"Marcus, you and Madge were impressive and looked splendid crossing the ballroom. You have made me very happy," said Margaret.

The year 1892 brought some excitement. Marcus Daly's horses made some inroads on the eastern circuit. "Montana" with Daly's jockey, Garrison, won the Brooklyn Handicap at 40-1 odds. Others in Daly's stable were winning numerous races. Still, Marcus was not noted among the eastern stables, which was fine with him, for it gave much better odds to his horses.

"Margaret," Marcus called from the doorway into the sitting room. "I've just come from the race track. Montana won the Brooklyn Handicap! I know everyone in Butte and Anaconda bet on Montana, so that should give them a little folding money. The odds were 14-1. It was wonderful, Margaret! The grandstand was full of unbelievers, and I enjoyed myself and made a little money on a side bet."

"Marcus, you haven't been drinking, have you? You are so noisy."

"No, me darlin', I'm just elated! Isn't this the greatest thing, Margaret? You know that Montana is an ornery creature. You can't tell what he'll do as you walk by. The stable boys are very leery of him, but he can run like the wind, and Snapper Garrison can handle him and bring him in first."

"Breeding shows, doesn't it? Ben Haggin told me that Montana would be a winner. I'm happy for you dear. Now settle down before you burst," said Margaret.

"I've come to have dinner with you and the children, Margaret. I find that I have to leave for a while and want to visit with my little ones, you, and my darlin' daughters."

"That's fine, Marcus. Where are you going?" said Margaret.

"I need to go to Butte and Anaconda." said Marcus.

"Marcus, you talk of your race horses, but how did the trotters do? You paid a great sum for the stallion, Prodigal."

"Well, Margaret, they are doing well on the midwest circuit. You remember that the son of Prodigal, Bitterroot won a big purse for me. Mr. J. Malcom Forbes and E. H Harrison bet against me on my trotter, China Silk, the two-year-old, a son of Prodigal. They put up $7,500 purse and I matched it. China Silk won in three straight heats on a half-mile track. I made a little on the side too, Margaret.

"You know Prodigal, Tammany and Hamburg probably will be my best stallions at stud. They have the right disposition, breeding, consistency and speed," said Marcus.

"Well, Marcus, I will admit I thought this horse business was a terrible expense, but it is starting to pay for itself, and I've never seen you enjoy yourself more than at the race track," said Margaret with a smile.

"Margaret, me darlin', I enjoy being with you on our own more than anything. I would give up all but you, Margaret. You're a great woman, darlin'," said Marcus.

"Oh dear, Marcus, you embarrass me."

Marcus laughed. It always amused him to see Margaret get flustered and blush. She never looked more beautiful, he always said.

"But, Marcus, you've not said why you need to go to Butte and Anaconda. Isn't the business going well?"

"Not to worry. The business is fine, but politics are getting very thick."

"Marcus, you're not going on with that dreadful feud with Clark, are you? We are related, since Miriam married his brother."

"Margaret, this is a man's business. I will not let William Clark smear my good name in his newspaper. I'm going to fight back, and he'll not hold a political office as long as I live. This is a war of words and I intend to establish a newspaper in Anaconda with a fine editor, John Dunstan. My paper, *Anaconda Standard*, will be the best newspaper west of the Mississippi!" said Marcus.

"Don't raise your voice to me, Marcus. I just think you and William could settle this without this slanderous and disgraceful behavior," said Margaret.

"I want this subject closed, Margaret. You know nothing of what that man has cost me in time and money with his ridiculous accusations."

With that, Marcus quickly left the room to find the children. They wouldn't upset him with this kind of talk, he thought. At dinner in the evening, Margaret and Marcus directed their conversation to the children, avoiding each other. After dinner was over and the children had gone to their rooms. Margaret and Marcus were forced to look at each other.

"Margaret, I don't want to start on a trip with you upset. It is a difference we have, but most of the time we agree. I'm very sorry I was curt with you and hope you can say the same to me."

"You are right, Marcus. I just don't like to see you hurt. This situation with Clark seems to get worse, not better," said Margaret.

"Come sit here beside me, Margaret, and we'll forget about everything for a while."

Margaret went to the divan and sat beside Marcus. As she sat down she opened the lid on the music box that played Vienna waltzes. The fireplace glowed from an ember fire, the music soothed and they sat in silence, close together, thinking their own thoughts.

Later, Marcus went to the Nederland hotel. He would leave early in the morning for Butte and Anaconda. John Dunstan would meet the train in Anaconda. John was a small man with curly sandy hair, a sharp nose and blue eyes that would have been steely but for their twinkle.

John had briefly met Marcus but had learned more about him from his fellow workers and acquaintances in Anaconda. He hoped Marcus would be as good to work for as he had heard.

The train pulled in and Marcus stepped off onto the platform. John waved and walked towards Marcus.

"Mr. Dunstan, thank you for meeting me. I'm anxious to talk to you. We can go to the Montana Hotel for lunch and talk business at the same time. Is that all right with you?"

At the Montana, Marcus sent the bellboy up to his room to ready it for his stay and to take care of the luggage. John and Marcus went into the dining room and were seated by the window looking out on the street at the passersby. Marcus wasted no time in getting to the subject at hand. "John, I want the finest newspaper possible. You are recommended as a fine editor. I want you to have a free hand and to make the *Standard of Anaconda* newspaper a reflection of your talents and dreams. I insist that William Clark's inflammatory statements about me be withdrawn, and I want an attack launched against Clark every time you can. I do not want him to be an office holder. I consider him unworthy, and now I believe him to be consumed with ambition, leaving his senses behind."

"We can do that. I'll do my best to make this paper newsworthy and trustworthy for the people," said John.

"Good. We understand each other, John. I will leave the paper in your hands."

Marcus decided to go to the Bitterroot and see about the horses and the lumber mill. He stayed in town at the house he had built for the family as a summer home before he had developed Riverside estate. This house contained offices for Jim Hamilton, some of the offices for the stock farm business, and bedrooms and baths on the second floor for visitors. Marcus had sent word ahead for Jim Hamilton to meet him at the railroad station. Marcus looked forward to being with Jim, for his reports were always up to date and assignments accomplished, and Jim was always eager to take on the next project.

"Jim, how is the irrigation project at the farm?"

"With what we dammed up on the Skalkaho and Sleeping Child Creeks, you can open up another 1,000 acres on the bench. The big ditch project will require more time and money. Taking the water from Lake Como requires some flumes being built on trestles to the river and also across the river to the east side. It is an ambitious project," said Jim.

"I know, Jim, but we want water on the east side for everyone in the valley, not just Riverside and the farms."

"We'll get it done, Marcus. It will just take a lot of planning and a lot of investors. The farmers and the orchard growers are coming into the office for information, and I can guarantee you they want the big ditch.

"Good work," said Marcus. "Say, Jim, can I have a rig? I want to go out to Riverside and the stock farm."

"Of course, take the one we came in. It's outside."

Marcus went out and got in the buggy, took the reins in hand and urged the horse on to Riverside. He passed the veterinarian's hospital and went to the thoroughbred farm and found Sam Lucas. "Sam, I'm here to find out about Tammany. Is he going to be able to do as well this year as he did last? I want to race him on the east coast."

"I'm telling you, Marcus. It's hard to believe that horse is a stallion. He has some puppies in his stall and he picks them up and puts them down with the most gentle mouth," said Sam.

"I know he is an amazing horse. Doesn't seem to have any temperament. I really love that horse. He is my favorite. He never lets me down. Where is he? I want to see him," said Marcus.

They found Tammany at the thoroughbred barn. When Marcus went up to Tammany, the horse whinnied and tossed his head.

"You remember old Marcus, don't you, Tammany? It was my lucky day when I got you for $2,500. I've paid a lot more for a horse, but none has proved any better than you. You are well-named; Tammany was the chief of the Lenye Indians, who drove the evil spirits from the Indian lands. He became such a legend that Tammany was adopted as symbol of resistance during the American Revolution and became Saint Tammany."

Sam said, "Do you think Tammany got that message, Marcus?"

Marcus laughed, "I think he did for he is a leader and knows it. I've had some men at the club ask me if I would race Tammany against the east coast favorite, Lamplighter. They've both won good purses. What do you think, Sam? I brought the times of Lamplighter. How do they compare to Tammany? I don't want to race him until he is ready. This race could establish the Daly racing stable as a viable enterprise. Up until now, everyone in the east think its a fluke and my ideas about feed, training and care are little eccentric."

"Let's see the times, Marcus."

Marcus pulled out the sheet of paper he had with him.

"Tammany has better times than some of these, but he gets better all the time. I think he could beat Lamplighter. Go ahead and set up the race. We'll get Tammany ready and Snapper Garrison will be aboard to guide him."

"Good, Sam. If Tammany wins this race, I'll build Tammany a castle."

Marcus returned to New York. The Hudson County Jockey Club offered a $5,000 purse for the winner. The race was to be in Guttenberg, New Jersey, on September 28, 1892.

On the day of the race between Lamplighter and Tammany the stands were full. 50,000 people were in attendance. The Marcus Daly family was in a box, and the backers of Lamplighter were in a number of boxes. People were there from all over the racing world.

The race started. Margaret tried to remain calm, she felt a hand clutch hers, and it was Marcus. Next thing she knew, she was to her feet waving her hand to urge Tammany on. Molly and Madge were hanging on to the rail of the box and leaning over to watch as Tammany turned at the top of the track to come down the home stretch. Mark and Harriot were jumping up and down calling, "Come on, Tammany! Come on, Tammany!" It was too close to call yet, but Snapper Garrison, not wanting to use his crop, was urging Tammany on, and he was responding. As Lamplighter made his last surge of speed, Snapper finally brought his crop up to tap Tammany; but, just as he was going to lay the crop on Tammany, the horse took off with a burst of speed and crossed the finish line winning the greatest race Marcus ever won, and the greatest race many ever had seen.

"Oh, Marcus, I think I'm going to faint," said Margaret.

"Not now, me girl, I don't have time to tend to you. Tammany needs me." Margaret had to laugh and the children thought it very funny and Madge and Molly giggled softly.

The odds had been 60-1 against Tammany. Marcus had made some side bets. Some say Marcus made $100,000 on the race, but it wasn't the money that was the thrill of that day. Marcus made his was way through the crowd to Tammany and Garrison and walked with them to the paddock as Sam Lucas joined them. "Marcus, was that a show — what a horse! I knew he could do it if he was handled right. Snapper did everything he should have to help Tammany win," said Sam.

"This is a highlight in my life, Sam. Thank you, and you, too, Snapper. Tammany, I will build you that castle I promised!" said Marcus.

Shortly after the race the entourage of Daly horses, trainers, veterinarians, supplies and horse paraphernalia was to leave for the Bitterroot Valley, home for the horses for the winter. The trip was delayed, and a terrible pall fell over the group. Tammany had pneumonia. Marcus was so upset and, as with the Irish men, anger relieves the pain of sorrow. Sam and Dr. Knowles were at the stall when Marcus arrived at the stable. "Where were you all after the race? Didn't Tammany get his rubdown and cool down after the race? Was he in a sweat without any relief? Where was everyone? Here is my greatest stallion, my favorite horse, and he has just won a race that will prove the Daly stables to the racing world. If that horse dies, you all go, including the stables. You had better work around the clock to save Tammany!"

Marcus was so agitated that he was sick. His face was flushed, and he started to cough. He turned and walked away, taking his handkerchief from his pocket and wiping his mouth and probably his eyes.

Dr. Knowles started rubbing Tammany with liniment and ordered cases of champagne. Tammany was hoisted in a sling so that his feet touched the ground but the main part of his weight was carried by the sling. Evidently Tammany tolerated the champagne treatment, for he did recover and a grand celebration followed. The leftover champagne was drunk in toasts to Tammany's good health.

Marcus joined in the toasts and said, "Doc, sorry I was so rough on you, but I couldn't believe that Tammany could have gotten pneumonia if he had had proper treatment."

"Marcus, it was my fault. I didn't stay and oversee his care after the race. It will never happen again. I'm completely humbled by this experience and am grateful that Tammany was saved!" said Dr. Knowles.

After Tammany was well, the horses were loaded into special cars with stable boys attending them. The three private railroad cars of Dalys were next. After loading the cars, they were hooked to the train that traveled across the cities and plains to Montana.

Marcus sent a letter to Jim Hamilton.

"Dear Jim,

As you know, Tammany won the contest against Lamplighter and the bookies are still working it out. It was a rout! We almost lost Tammany to pneumonia, but he is recovered and on his way to Riverside.

That horse is my pride, and I told everyone, including Tammany that I would build him a castle if he won against Lamplighter. Another project for you, Jim — draw up some plans and send them to me. I want to get started on construction as soon as possible.

Tammany's Castle specifications are as follows. The building site I have chosen is on the hill between the manager's house and Sam Lucas' house. Since it will be unobstructed by the other buildings, I want both the front and back of the stable equal architecturally. The exterior I want red brick, with cement walkways and landscaping as fine as our homes.

Inside I want a tack room with green leather furnishings and display cases for trophies, ribbons, the thoroughbred horses lineage and files. I want it to be a place where a buyer and myself can have a comfortable spot to research and discuss horses and prices.

I want fifteen stalls heated and padded. I want cork floors throughout the building. On the stalls I want the name of each horse and his record of wins and monetary rewards.

Outside I want a covered walkway with arches across the front so that the visitors can get from the carriages to Tammany's castle when there is inclement weather. And, Jim, I've seen the barns in the east, and also some of the homes with cupolas, topped with weather vanes. I like that. I hope my specifications can be incorporated into an outstanding architectural building.

I know you are busy overseeing the electrical plant being constructed for Hamilton. We planned for the future there. The plant can supply a city of 10,000. Better to have an excess than adding on later, right?

I've got the nursery planting 13,000 poplar trees along the lanes on the seven farms at Riverside.

Thanks for taking this project for me, Jim. I want

Tammany to have the best.

I'll be in Hamilton in February this year. I need to stay here in January for Anaconda business.

Sincerely yours,
Marcus Daly"

When Jim received the letter he called his staff together. "Men, here is a project that should bring a bit of the artist out. Mr. Daly wants a castle built for Tammany."

Jim read the specifications and the draftsmen and engineers took them down.

"Let's see what we can come up with," said Jim.

It was December 1893 and Marcus had another debutante to present, Molly. He was leaving the hotel to be with the family and to be briefed on the debutante ball, and would be glad to have this one over.

After dinner Marcus, Margaret, Madge and Molly sat in the living room and made their plans.

"Molly, tell your father about your dress and your escort," said Margaret.

"Papa, I have a beautiful Worth gown from Paris. I loved Madge's debutante dress, so I wanted as lovely a gown. It's white, of course. We all wear white."

"I know that, darlin'," said Marcus.

"Well, anyhow, mine is white lace with a large flounce of lace embroidered with white flowers and pearls centered around the neckline and shoulders. The bodice is tight fitting and an extra embroidered lace piece falls along the side."

"Why don't you show me, Molly?" said Marcus.

"Oh, Papa, it doesn't look like anything when no one is in

it. You'll see it the night of the ball," said Molly.

"All right, how about your escort after you have been presented?" said Marcus.

"Well, I've been to a number of parties where this one friend has been most attentive. I like him, but I'm not ready for any commitment. Please be kind to him, Papa. His name is Robert Van Holt. He is tall with blonde hair, blue eyes and is going to Yale where he rows on the team. We have fun dancing together and laugh at the same things."

"I'll be kind for you, Molly. I won't embarrass you!" said Marcus.

Madge added her advice to the conversation.

"Molly," said Madge, "Papa was very comforting to have beside me as I walked across the ballroom. The room looked as if it was a mile long."

"I'm so glad you're around to advise me. It must have been hard for you to be on your own without a friend to tell you what to expect. Of course, Mrs. Cornelius helped you, and me, too, but it's better to have you tell me."

The conversation between the family was spirited and happy. The ball was the next evening.

"Molly and Madge, don't talk all night. Molly, you need lots of rest so that you will be at your best," said Margaret as the girls left the room to retire.

"Promise, Mama, I want to get to sleep so that I can get rid of these butterflies in my stomach," said Molly.

When Molly and Madge had left, Marcus turned to Margaret and put his arm around her and drew her over to him. Even with the tight corset and all the clothing, Marcus felt a surge of passion.

"Come, my girl, let's go into the sitting room in our bedroom so that Cutler can ready the living room for tomorrow. No sense keeping him up any longer," said Marcus.

"Marcus, are you spending the night here — is that what you're leading up to?"

"Ah, me darlin', I can't fool you. If it is all right with you, Margaret, I would like that," said Marcus.

"Well, I suppose, Marcus."

Margaret always acted as if this was a burden she bore, but really she loved Marcus and their nights together were reassuring to each of them. Their love had survived separation and different interests.

Marcus and Molly were presented at the ball. It was not as hard for Marcus to walk with Molly as it had been with Madge. The second time around eliminated the unknown, and Molly was more assured than Madge, for Madge had not left out any detail of her experience as a debutante, which helped Molly be more confident.

"Papa," said Molly, "here we are at a milestone in my life. I'm glad you're here, Papa."

"And where else would I be but with my darlin' daughter, who is in need for me to present her proudly to New York society. You are a beautiful child, Molly. I'm very proud of you. Beside, darlin', you are clever at your school work. Beauty and brains, just like your mother."

Molly blushed and said, "Thank you, Papa, I'm so glad I please you."

Molly needn't have worried about Marcus meeting Robert Van Holt. Both were fond of sports and actually Molly got tired of listening to them talk.

"Robert, in Montana I have a basketball team that I send around the country playing colleges. They always win," said Marcus.

Molly interrupted, "Papa, you have told Robert about horses, football and basketball, and now Robert needs to pay attention to me and to his dance card."

Marcus and Robert laughed and went their respective ways. "You and Molly have a good time, Robert. I'll look forward to our meeting again. Maybe you can come see my horses race at Belmont," said Marcus.

"I would like that, sir. It has been my pleasure to meet you," said Robert. Marcus returned to Margaret and together they watched the blur of white dresses and black tail coats whirl around the dance floor.

. . . .

Christmas came with the family at home at 725 Fifth Avenue. The same customs were observed each year: The first one up wakened the rest of the household to unwrap the gifts that were under the tree. Harriot vowed she would be the first that year. Mark had always beat her, but she planned to go to sleep early so that she would get up first.

"Mark, I'm going to be the first one up this year. I have a plan to beat you. You know if you get up before 5 a.m. it doesn't count," said Harriot.

"You won't win, Harriot. You're always in bed too late to wake before me," said Mark.

"I know, but this time I'm going to," said Harriot.

Harriot went to her papa that Christmas Eve.

"Papa, you have a clock that rings to wake you. Could I have it for tonight?" said Harriot.

"I suppose so. Are you trying to awake before Mark tomorrow morning?" asked Marcus.

"Yes, Papa. He's always first and I can't seem to wake up before he does. I think your clock would do that for me," said Harriot.

"Well, my girl, just so it isn't before 5 o'clock in the morning," said Marcus.

Harriot went off with the new invention, the alarm clock.

Harriot won, Christmas came and went, and 1894 was welcomed as the New Year at 12:00 midnight.

After the win of Tammany over Lamplighter, the odds against the Daly horses were not as great. The Daly stable of horses was holding its own, bringing money into the enterprise and validating Marcus' theories on training and care of thoroughbreds in the high altitude of Montana.

Jim Hamilton had sent blueprints and elevations of Tammany's castle to Marcus. He chose the red brick building with gable roof, white cement arches and two cupolas with weather vanes. The interior wood and stalls were mahogany instead of pine. The stable known as Tammany's Castle was finished by the time the Daly family arrived at Riverside for the annual summer reunion of family and friends.

After the family had settled into their rooms at Riverside, Marcus went to Margaret and said, "Margaret, do you want to take a little ride over to see Tammany's Castle?"

"Yes, Marcus, I would. I thought the plans were very nice and would like to see the finished product," said Margaret.

"I'll call for a buggy and one of the carriage horses to take us over. 'Tis such a nice day and I'm so anxious to see the place," said Marcus.

As they came up the hill past the mansion they saw Tammany's Castle for the first time. It looked as if it belonged to Churchill Downs, with its double cupola.

"My goodness, Marcus, that is a castle. I'm sure the horses will be comfortable there," said Margaret.

"Aye, and they should be. They bring me much happiness and soothe me," said Marcus.

The landscaping around the building was meticulous. Grass was planted and was spring green around the entire building. A cement sidewalk led to the front entrance. The sidewalk parted around a flower bed and joined itself again and continued on to the front entrance. There were shrubs on each side of the five arches which were across the building. The back of the building was the same with the same landscaping.

Tammany's Castle not only was beautiful from the outside, but also inside. The specifications had been followed. The brass trim was bright from polishing, and the floors were soft to the hooves of horses and the feet of people. The brick walls inside had been rounded on the corners so that the horses wouldn't be cut if they should happen to turn the corner too soon.

"Marcus, I'm impressed. This is an architectural masterpiece -- to think it's a stable!" said Margaret.

"Jim, come here. Mrs. Daly and I want to thank you for Tammany's Castle. No one could have done it better," said Marcus.

"Tell you the truth, Marcus, this was a project that was a real pleasure. Everything went into place with your plans and my execution of the work. It is a showplace," said Jim.

After Margaret and Marcus had toured the facility, they walked out to the buggy and started the journey to Riverside.

"Margaret, you know how everybody in Anaconda favored Tammany. Well, I'm going to have that artist in New York that does pieces of wood into a portrait do Tammany. He hasn't done an animal yet, just people, but I'll pay the same for Tammany's portrait as he charges for people, and I'll have it embedded in the floor at the Montana Hotel in the bar. What do you think of that, Margaret?" said Marcus.

"Aren't you afraid they will walk on Tammany's portrait, Marcus?" asked Margaret.

"No, and they had better not when I'm around," said Marcus.

The next day Marcus sent a letter to the artist and a picture of Tammany's head for accuracy. Within two months the portrait was shipped and installed at the designated place in the saloon in the Montana Hotel in Anaconda.

Marcus was right, no one walked on the portrait. Some even bowed to Tammany as they went past his image. The miners and townspeople had won a goodly sum of money betting on Tammany!

There was an article in the Hamilton monthly paper that published the biggest taxpayers in the county. Marcus Daly led the list, paying one-sixth of all the taxes.

Marcus brought the paper to Margaret to show her. "Margaret, look at the head of the list in this paper," said Marcus.

"I'm not surprised, Marcus. There is no other operation as big as yours. By the way, dear, I appreciate your giving land to all the churches in Hamilton. That is very generous of you. So many could not buy land and build their church, too. We want this to be a good Christian town, and I believe it is. Most of the denominations have claimed the offer of land, haven't they?"

"Yes, Margaret, they have, and I will help with building if need be. With the tax base the county is trying to establish schools, and I told them that I would supplement the school fund when needed. After the valley is established they will have enough tax funds to become self-sufficient."

"Well, Marcus, no one can say you're not generous, and Hamilton will be your legacy. Not many towns are planned by an architect engineer such as Jim Hamilton or have the resources to build as substantial a base as Anaconda Copper Company and you have given Hamilton. Let them call it "only a company town." Someday it will stand on its own, thanks to your support," said Margaret.

"That is quite a speech, Margaret. I didn't know you felt that way. Those are kind words for this old Marcus. I get lots of criticism from other quarters. Thank you, Margaret."

Chapter Ten

The years fell into a routine: Fall meant horse racing and children in school; at Christmas the family gathered; in January Margaret and children went off to Paris, and Marcus to Montana; spring brought more horse racing, and Margaret returned to New York; and summer brought everyone back together at Riverside.

This year of 1894 brought a special celebration to the town of Hamilton ,for it finally was incorporated. The stables kept increasing and more help was needed for the horses. That summer eight Negro boys between ages eight to fifteen years arrived in Hamilton. Marcus went to the stable to see Sam about the schedule of horses going and coming. As he came up to the stable he saw the new young men getting out of the wagon. Sam came out of the stable and ran into Marcus. "Sorry Marcus, I didn't see you. I was going to see about the boys," said Sam.

"Aren't these boys young to be working?" asked Marcus.

"Well, if we can get them young they are easier to train. They will work with the horses and we will train them as jockeys," said Sam.

"What about school, Sam? They can't go to public school and work out with the horses," said Marcus.

"I know, Marcus. I have a tutor for them, and they will go to school four hours a day," said Sam.

"I guess that will be all right. Their extra activities will be horses instead of sports or glee club, eh, Sam?"

Sam nodded and smiled. "Right, and Marcus, and I'm having uniforms made for them, since they only have the clothes on their backs."

"Be good to them, Sam. I remember myself how hard it

was to leave home and family at such a young age," said Marcus.

"I will. They will be some of the lucky ones," said Sam.

"I'm going to Butte tomorrow, but will be back by the end of the week," said Marcus.

"Butte gets wilder each time I go. There's much dissension among the miners, The different nationalities working the mines don't assimilate and there are distrusts and fights among the Irish, Norwegians, Cornish, Finnish, and the rest. I hate to go over there since we lost the fight for Anaconda to be the State capitol. I kept up with Clark's soliciting votes. Money and favors for the votes is a rotten way to get votes, but if one side does it, the other has to follow. It's the same in the east. Then there's Heinze. He's a rascal, which is a better word for him than he deserves. Now he claims the apex of the Anaconda Mine and other large mines are on the small parcel of land he purchased by searching records of unclaimed land. Can't be over an acre. It's ridiculous, but I suppose we will need to hire more lawyers to fight his latest shenanigans. This is way out of your interest and responsibility, Sam, but Anaconda is what pays for the stable. So, when you vote, vote right, Sam, and tell anyone who will listen," said Marcus. "I'm sponsoring the Democratic Party in New York with some goodly amounts of money. I hope to be of some influence in the national politics in this country."

"Well, Marcus, I wish you luck. I'm glad it is you, not me. I wouldn't know how to cope. If it's horseflesh or a race, I'm competent," said Sam.

Marcus arrived in Butte and took a look around at the smoke puffing out of stacks, mine tailings making new barren mountains and the wild and quarrelsome population. This is a good thing that the wind blows enough to dissipate the smoke, he thought. Something had to happen to straighten Butte out from this rough existence. Many people of means and character left Butte for other places, as they could not feel safe in that wild and tough town.

George Hearst had died in 1891, a shock to Marcus. Arvid Simmons, a representative of Mrs. Hearst and an old friend of Marcus, came to Butte in the spring of 1895 to discuss the Hearst Estates holdings in Anaconda Copper Company and the possible sale to Standard Oil Company. Marcus met with him in Anaconda at the Montana Hotel.

"Simmons, it has been some time since you've been out here. Glad you made the trip for I have a few items to discuss with you," said Marcus.

"I have a agenda for you, too, Marcus. We've got to decide which way we are going in Anaconda Copper," said Simmons.

"Right, but before we get into where we're going, let's clean up where we are," said Marcus. "That crazy Heinze has found a piece of property at the edge of the Anaconda mine and two other mines. He claims the apex for all the mines are on his property. It sounds crazy and it is crazy, but whenever we go to court to have his claim thrown out, that judge Heinze has in his pocket, rules for Heinze. We can't get anywhere, and it's costing millions to defend our cause. Besides, he's sinking shafts and taking ore from other mines. He is not a small time operator. In his speeches to the miners and his rabble-rousing newspaper, he rallies them to his cause. I'm telling you, Simmons, he has driven all the mine owners crazy. One thing you had better do, is buy up all the worthless land around Hearst's Homestake Mine. If Heinze can do such damage here, someone could do it in South Dakota," said Marcus.

"Ye gods, Marcus! I knew there was a hassle here, but I had no idea it was so disastrous. I'll be on the wireless today with instructions to buy the land around the Homestake," said Simmons.

"Good. It's better that you spend a little to guard against losing a lot."

"Marcus, with all this hell going on here in Butte -- with it costing so many hours of work, creating discontent among the miners, and costing more money than is imaginable — let's sell now to Standard Oil Company. They've got the money. They're a public stock. You can retain enough power to run Anaconda without all the pressure. Surely, Heinze can't defeat Standard Oil Company and all its resources," said Simmons.

"Simmons, it would tear me up to sell out to Standard Oil. Probably I'm selfish, but I built this company, and George Hearst, Haggin and Tevis have given me complete control and their confidence in 90% of company matters. I hate to give that up. But, if the Hearst interest, Haggin and Tevis say sell, that's the way it will be," said Marcus.

"Don't look so sad, Marcus. You know you will still be running Anaconda. Hell, Standard Oil hasn't anybody that could take your place. They'll probably bump you upstairs so that you'll be around while they train someone to take your place here in Butte. Want to bet on that one, Marcus?" said Simmons.

"I know you're right, Simmons. I can't take care of everything plus what we keep acquiring, forever," said Marcus. "So I guess the dye is cast. Standard Oil Company is going to get Anaconda. I've fought it as long as I can," said Marcus.

"Marcus, you'll be twice as rich and have half the work. It has to be better all the way around," said Arvid.

Marcus didn't know how he felt. He knew Simmons was right about less responsibility and work, but it was difficult to give up the control.

"Well," said Marcus, "I'll assume this move to Standard Oil will be for the best. I'm glad that it's decided. Now that it's summertime I can go to Riverside and forget this nightmare of the mines, miners and mine owners. Margaret is opening the mansion at Riverside this week. Will you come with me and stay a bit while I show you our lumber operation for Anaconda and my personal operation at Riverside?"

"You know, Marcus, I'm going to take you up on that invitation. I need a little relaxation, too. Besides, it will be a good time for you and me to talk about the good old days," said Simmons as he gave Marcus a slap on the back.

"Good! Let's go drink to the good old days and wind up our business here by the weekend. I'll send word to Margaret that we'll be there Saturday. She'll be glad to see you."

True to their promise, Marcus and George arrived at Riverside on Saturday afternoon. Eddie was there to meet the men with the Victoria carriage, as it was the most comfortable to ride in and had room for luggage to be put up with the driver. Eddie greeted Mr. Daly and bowed to Mr. Simmons, welcoming them home. He put their luggage on the front seat and floor by him, the men climbed into the Victoria from each side, and they were on their way. Young trees were planted along the way, the grass was green, the timothy hay was up about eight inches, the sun was setting, and the mountains were blue with daylight failing.

"What a gorgeous time of day, Marcus. And what a view! I had forgotten about these Bitterroot Mountains. It's been a while since you went searching for George and found him not far from here."

"That was a happy moment in both our lives. George knew he was all right, but Mrs. Hearst and I weren't sure. 'Twas a great relief to see him that day," said Marcus.

Margaret was downstairs when the men came in. Hearing their voices,she came to the north portico to greet her husband and guest.

"Margaret," said Mr. Simmons, "you look just like the young girl I last saw. Time goes by and it doesn't touch you. What is your secret?" said Arvid.

"Well, Arvid, you're still the same. You're not even Irish and you've kissed the Blarney Stone, but I thank you. It is so good to have you here. How could we let such a long time pass? How are Helga and Mrs. Hearst?"

"They're fine, but very busy with civic opera, libraries, museums and such. Mrs. Hearst is doing more than her share to bring culture to San Francisco and Helga has been on a number of committees with her."

"That's wonderful. We women have to civilize while you men have your life's work and interests. Hamilton's opera house will open this summer, and they're building a hotel which will have a ballroom," said Margaret.

"Really, Hamilton is very civilized," said Arvid.

"Yes, it is. We'll go into Hamilton tomorrow so you can see the lumber operation and the town," said Marcus. "But before we do anything, you must see the seven farms of Riverside."

"That's what I came for. That and to have a visit with you and Margaret."

During his stay, Arvid met the family and guests staying at Riverside, he saw the farms, Hamilton, the sawmill, the dam, the raw timber crashing into the river, the new orchards being planted, and the irrigation system. He saw it all.

"Marcus, this has been a great experience. I knew about your racing stable, but I had no idea what had transpired here in the last five to six years," said Arvid.

"Well, we've been busy," said Marcus

Arvid laughed. "Busy? Yes, I think you must have been hysterically busy. This has been a herculean task!"

"Well, now, a little praise is appreciated. We have good men working for us in the Anaconda Company, and I have good men taking care of my enterprise here in the valley," said Marcus.

When Arvid had left and Marcus had came back to the mansion, he sat down in his office and started to read the reports. Margaret rapped at the office door and Marcus rose. "Come in, Margaret. Thank you for being such a good hostess. Arvid did enjoy himself and you had a lot to do with that, darlin'. You made him feel welcome and he did relax."

"Thank you, Marcus. I've always enjoyed Mr. Simmons and it was a pleasure to have him here. I don't know how much relaxing he did. You two were all over the countryside," said Margaret.

"We had a great time."

"What are you doing, Marcus?"

"I'm shuffling papers, darlin', and not getting anywhere. I was looking at the horse wins and balancing out with costs. We should break even in a few years -- maybe two.

"Are you gathering your sisters and guests for the fair? This will be the first Ravalli County Fair. I think it will be successful, and it'll be wise to join the Fruit Fair in Stevensville with our horse races."

"Yes, dear, we're all going the first day and will return off and on during the two weeks. I won't even ask you. I know you will be there the entire time," said Margaret. "We'll be having tea soon. Are you going to join us?"

"No, darlin', I have all this reading to do, then I must get some letters out. I didn't work at all while Simmons was here. I'll be there for cocktails at 6:00."

The Ravalli County Fair was a great success. Ten thousand people came the first few days and the crowds were very adequate through the two-week fair.

Marcus went over the horse wins and, although there were some by Tammany in 1892, there hadn't been many wins the next three years. He talked to Sam about this and he told Marcus that Odgen was eligible to race in the next season on the east coast. He thought Odgen was a great hope. Sam didn't want Marcus to be discouraged, but Marcus was. Marcus wasn't going to disband the stable, but maybe pull back some. He would see how Odgen did this next year, 1896. For a brief minute Marcus was depressed. He allowed himself just a moment to feel sorry for himself, then plunged back into the work at hand. He'd received a letter from the Democratic Party in New York City.

Dear Mr. Daly,

You have been a fine supporter of our democratic candidates for national offices. I want to call to your attention that the famous lawyer and orator William Jennings Bryan is likely to be nominated by the Democrats to run for President.

As you know, his main platform is getting off the Gold Standard and into the Silver Standard. As our country grows, we need more funds. Gold is scarce and expensive whereas silver has a lesser value but it is more plentiful.

We would appreciate your support of Mr. William Jennings Bryan in any way possible. We are taking monetary donations and, as soon as he is assured the Democratic party nomination for President, we'll put this money to good use in our effort to win the Presidency.

Sincerely yours,
Mr. John Morgan
Democratic Party Chairman"

Marcus immediately answered this letter.

"Dear Sir,

I am very willing to support Mr. William Jennings Bryan for President. I agree with his policies and admire him as an individual.

Should he be the Democratic Parties nominee for President, I will donate $250,000 to his campaign. I would also request that he make a trip to Montana as my personal guest. I will have him speak in Butte and Anaconda and in Hamilton while staying as a guest at my home at Riverside.

Please advise me of this possibility. It would be an honor, and the trip and expenses would not come from his campaign funds. He would be my personal guest.

Sincerely yours,
Marcus Daly"

The horses took a back seat while Marcus became involved in William Jennings Bryan's campaign.

William Jennings Bryan succeeded in his nomination as the 1896 Democratic candidate for President. Marcus' plan for Candidate Bryan was received with great enthusiasm. The family and Marcus were at Riverside when they heard the news.

Marcus knew Margaret was taking her three o'clock horseback ride and would be back shortly. He pressed the buzzer on the side of his office door to call for Cutler.

"Mr. Daly, your rang?"

"Yes, Cutler. When Mrs. Daly comes in from her ride, please tell her to come immediately to my office."

"Yes, sir, Mr. Daly, I will do that," said Cutler.

"And, Cutler, please bring me my Irish whiskey and water here in the office. I'd like a little snack with it please -- something from the tea tray."

When Margaret came into the office she had a stricken look on her face.

"Marcus, what is the urgency? Are you all right?"

"Aye, I am, my darlin'. We are going to have our most important house guest and an honor it will be! William Jennings Bryan will be here in August during the fair and will speak about all of his plans for our country should he be elected to serve as our President," said Marcus. "I want everything to be perfect for him. He is a man I very much admire. I especially admire his oratorical talent. I do want him to win the election."

"That is very nice, Marcus. I am honored to have him as a guest. We'll all enjoy meeting him and will try to make his visit pleasant," said Margaret.

"I know you will, me dear. I just wanted to give you as much notice as possible. I just got the word today. You only have two weeks to plan. Thank goodness the Ravalli Hotel is finished, many people will be coming to the fair and to hear Mr. Bryan. Hamilton will be able to show Mr. Bryan what a fine community it is, having changed from a logging boom town into a respectable community with a solid economy rooted in agriculture and lumber."

"Well, not many towns can be planned from the start, Marcus. It does make the town unique." said Margaret. "I did go to the hotel with Hattie and Mattie. We were given the grand tour of a few of the rooms. There are electric chandeliers in each room and hot and cold running water in the marble wash basins. The hotel even has call bells and steam-heated rooms. The manager insisted that we have lunch in the dining room, and the food was very good. The dining room must seat 60 to 75 people, and it was almost full. That shows how much it was needed for it has just opened its doors. When the carriage came for us the manager walked with us out the door. 'My dear ladies,' he said, 'see this building to your left here? That is the laundry and the servants' quarters. That was Mr. Daly's requirement. The servants' rooms are very adequate, and certainly the building is impressive isn't it? I so like that the building matches the hotel's red brick and has a fine front entrance.' He was proud of the hotel and environs."

It was time for the great man to arrive in Hamilton. Marcus had gone to New York with his private car and returned to Butte with William Jennings Bryan, who was to give one of his famous speeches.

By the time Marcus and Mr. Bryan arrived in Hamilton,

the fair was underway and the speaker's stand was decorated and ready. The hotel was full, and the crowds were the largest ever.

This was Democratic party country, and the cheers for their hero, this learned man, were thunderous. Both Mr. Bryan and Marcus were stimulated by the tremendous reception to his words. Nevertheless, the night before his speech, William Jennings Bryan was happy to be away from the crowds, relaxing at the Daly mansion with the close friends and family of his friend and supporter, Marcus Daly. Marcus would allow Mr. Bryan his rest before his speech, but he had urged Mr. Bryan to stay an extra day to tour Riverside's seven farms, especially the thoroughbred operation and to see the lumber mill.

"Mr. Bryan," said Marcus, "this visit of yours is a first for Montana. A candidate for President visiting our state. We are honored. I want to propose a toast to your success. You are the most qualified to lead our nation."

The day of the speech Marcus sat on the speaker's stand with Mr. Bryan along with the president of Ravalli County Bank, the manager of the mill, the manager of the company's mercantile, and Anaconda Company's business lawyer.

Marcus wanted Jim Hamilton to be there, but Jim had left town on important business, or so he said.

William Jennings Bryan gave a stirring speech and the crowds were mesmerized by his eloquence. Afterwards, Marcus and the others escorted Mr. Bryan through the crowd into the stands to watch the horse races.

"Mr. Bryan, these are some of my horses racing here. I never advise anyone on how to bet, but here is a sheet telling their breeding and their times. I'm going down to bet if you care to join me," said Marcus.

"I think I will. I can't resist betting on a horse that has that certain spark. I saw one out there and I see by its number it is one from your stable."

And the horse won for Mr. Bryan. "Say, you might think about pursuing a career in horse racing," said Marcus.

The next day they toured the facilities in Hamilton and at Riverside Farms. Then Mr. Bryan went to the east coast, campaigning all the way to New York.

Marcus went to see Sam and again wondered if any of the horses were eligible for the eastern circuit. They had barely made expenses the last three years.

"Sam, what do you think about this record at the race tracks this year? Am I wrong on my theory that the best of breeding and environment produce winners?"

"No, Mr. Daly, it just takes time. You had some phenomenal results early and the pace slackened. We have some great horses coming along. Your theory is right. You've made a believer out of me," said Sam.

"All right, then, I'll leave the decisions up to you. I'm going to New York to help Mr. Bryan's campaign. He was a great congressman, I like his ideas, and I'll do all I can to back him in person and financially," said Marcus. "Then I'll be negotiating with Standard Oil these next few years on the buyout of Anaconda Copper Company. I know you've heard rumors of this, and it's true. It won't affect you -- I'm not running this stable as a company project. This is my stable. I'll probably be wealthier if Standard Oil takes over than I would be if they didn't. They'll give two shares of stock for the price of one, based on my estimates of how much copper is still to be mined."

Marcus laughed. "Sam, I don't suppose you give a hoot about all this Anaconda business, but sometimes it helps me straighten out problems by discussing them. Unfortunately, I really don't have too many disinterested persons that I can talk to. You are one of them, Sam, and I thank you for letting me natter on," said Marcus.

"I'm honored you want to talk to me, and I understand what you're saying. Sometimes it's easier to discuss a problem with a stranger or an uninvolved bystander than with a close associate," said Sam.

"Well, Sam, I'll see you in the spring of 1897 at the farm in Maryland. Take good care of my horses. Let's bring in some winners!" said Marcus.

Marcus returned to New York with the family. Mark and Harriot were growing up, and Molly and Madge had been introduced to society and enjoyed their piano, reading, and friends, but had not found the right combination for life partners. They were not unpopular. Money talked, and New York now knew that the Dalys were very rich and because of their pleasant demeanor were beginning to have good friends and to be accepted in the social circles of New York.

The elections were in November, the first Tuesday of the month. With such a diversified group supporting Bryan, Marcus was sure he would win. Bryan was supported by the Democratic party, The Peoples Party and the Free Silver Republicans, he was defeated by the Republican candidate, William McKinley.

Marcus was not a good looser. "Margaret," he called, "can you believe Bryan lost the election? He is such a fine man. He would have made a great President."

"Marcus, calm down. I know it's hard to have your friend lose, but you did as much as you could and that should be a comfort to you. You know there is another election in four years, perhaps he will win then," said Margaret.

"That doesn't make me feel any better. I'm completely disgusted," said Marcus.

Margaret listened to Marcus for another half hour until he had finally had vented his anger. Molly and Madge came to the door during ravings, but Margaret waved them away.

Finally, Marcus sat quietly and Margaret took out her embroidery. She lifted the lid on the music box. The music was soothing. Margaret started to say something to Marcus, but he had fallen asleep. How amazing, thought Margaret, I would have been awake for hours if I had been that upset. Margaret smiled and reached for a lap robe to put it over Marcus. It was better to let him sleep than to wake him to go to bed.

Later that night, she heard Marcus getting into bed. Marcus had been very quiet, for he didn't want to waken Margaret.

"Marcus, are you all right now?" said Margaret.

"Yes, me darlin'. Sorry I was so loud and boisterous. I must have hurt your ears."

"I understand, Marcus. It's always hard to lose when you believe strongly."

"Thank you, darlin'. I'll see you in the morning."

The year 1896 was ending with another Christmas of joy at the Daly household.

"Margaret, are you going to Paris this year?" said Marcus.

"The girls and I will go for a month, but Mark and Harriot

will be here. They're well established in their schools and have made friends. Harriot and Mark will go next year. We'll stay three months, then enroll them in school. I think one year in Paris for them is enough. So much is offered at Miss Porter's school for girls and at Choate that a foreign education isn't as important now. It is important that they speak French fluently so the three months of intensive training should complete their French language. Why do you ask, Marcus?"

"I just wondered. Maybe I'll join you for awhile. We'll see. It depends on Anaconda and Standard Oil."

Marcus went to the Maryland farm in March to see his horses and Sam.

"What are the odds now, Sam -- are we going to have a few big wins? Thank God, Ogden won the Belmont Futurity or we would be in the red," said Marcus.

"Mr. Daly, if you could buy this greatest of horses, Hamburg, it would be a wonderful addition to your stable," said Sam.

"Why can't we buy it, Sam? I told you I wanted the best."

"John Madden has Hamburg and he won't sell to you, but Billy Lakeland will buy Hamburg and sell him to you for $41,000."

"Sold," said Marcus.

And the deed was done. Hamburg was the greatest of stallions, and with this purchase Marcus had a renewed interest in the Daly stable.

The spring racing season was a success and Scottish Chieftain won the Belmont Stakes.

Marcus decided to have his horses portraits painted. Tammany would be first. Marcus had retired Tammany to stud after the great Lamplighter vs Tammany race. He commissioned artist Harry Stull to do Odgen, Montana and Tammany. Every muscle and every bone showed in their lifelike portraits. Tammany's showed his beautiful light chestnut color with the white blaze on his nose. The others were just as well done. The four-foot by five-foot paintings were framed in gold-encrusted frames, and Marcus had them hung in the Ravalli Hotel in Hamilton. The others were to be put in the Montana Hotel in Anaconda.

Marcus heard of Henry H. Cross, who painted scenes on circus wagons for Barnum and Baily. Marcus wanted a landscape with his finest horses grazing. Cross painted four paintings of a size that could pull down a wall -- each of the paintings were eight feet tall and eleven feet long. One scene included the Bitterroot mountains in the background of a lane with the poplar trees lining the drive. Marcus had them hung in Tammany's Castle where the posts and beams were strong enough to support the huge paintings.

Before the Dalys had left the Bitterroot Valley in 1896, Margaret decided to have the house remodeled. The Queen Anne Victorian style was more popular now and the finished remodeling would be done for the summer of 1897. Margaret was most anxious to return to Riverside that summer. She was sure that all would be well with the house, but she still worried about the project.

"Marcus, as soon as Mark and Harriot are through with their classes in May, I would like to go to Riverside. I know you won't be able to leave that soon, but I'm so concerned about the remodeling. I'm sure everything is fine, but I'd like to be sure. If it isn't finished, we can stay at the hotel for a few weeks."

"Of course, Margaret, you can do whatever you want. I'll be busy here until June, but I will join you then." said Marcus.

"The irrigation system is being completed. We have the Hedge Ditch, which is about eighteen miles long; and I purchased the Republican Ditch and added a canal from the river. This will irrigate any of the acreage on the Riverside farms. And Jim Hamilton is still working on plans and I'm gaining some support for the big ditch from Lake Como. Then the valley will be fruitful. Irrigation water is the most important ingredient for successful farming in the Bitterroot."

When Margaret and the family arrived they found the Queen Anne remodeling of the mansion was successful. It looked like a big wedding cake and the cranberry glass in the turret windows looked like the flames of the candles on a cake. Everyone loved it and thought it much prettier than the original Victorian. Margaret waited for Marcus and the guests to arrive. She was anxious to hear their remarks about the remodeling. Marcus would be in the next day if all had gone well. He did arrive on schedule and as he was driven up the lane he began to catch glimpses of the house.

Margaret was waiting for Marcus on the north portico. Marcus, dear how are you? We've been waiting for you to start our summer activities. Of course, we've been riding and playing croquet but haven't even been into Hamilton," said Margaret.

"Well darlin', I've been trying to get here, but between horses and Anaconda, I'm just happy to be here," said Marcus.

Margaret followed Marcus up the stairs. Cutler took care of the luggage. Marcus went into his room and Margaret was still behind him.

"What are you doing?" asked Margaret.

"I'm going to get cleaned up," said Marcus.

"I want to know how you like the remodeling of the house, Marcus," said Margaret.

"From what I've seen I like the inside, but it looks like a church outside. It doesn't look like a house to me." said Marcus.

"Oh dear, Marcus, that's dreadful. You don't like it. What can we do? I know how you love coming here and I don't want anything to mar your happiness," said Margaret. Tears came to her eyes.

"Now, Margaret, don't cry. Maybe I'll get used to it. If I don't we'll go to Mr. Gibson in Missoula and have a new face-lift on the mansion. Don't fret, Margaret. Maybe tomorrow it will look fine."

"Come here darlin', come stay with me for a while. Its been a long time since we've been together," said Marcus.

Marcus closed the door and then two had their time together to comfort each other and for a time to close out the world.

The next two days the guests arrived and life at Riverside took up in earnest. With the familiar and comforting routine, friends and family gathered in friendship and love in their time for escape into the world of summer at Riverside.

Marcus went into his office, as was his usual custom, at 10 a.m. He decided to list some of the winning races of his horses. Tammany had won three big races at Belmont Park, Odgen had won a number of good purses, and Hamburg would be running this year and had a good chance of winning some good purses. Marcus hoped that the stables were starting to make money.

September came and all the goodbyes and thank-yous tearfully ended. The group dispersed to relive the summer at their homes in California, Butte, Anaconda and New York. The best part was that they knew they would meet again th enext year at the same time and the same place. In the meantime, letters would bring the news of each other.

Marcus came home from his office in New York and joined Margaret, Molly and Madge for tea.

"Here are my lovely lassies. Three pretty ladies, may I join you? I have something to tell and something to ask," said Marcus.

"Of course, Marcus, come in," said Margaret.

"Oh yes, Papa, do join us. We haven't been with you for ages," said Madge.

"What do you have to tell and ask us, Papa?" said Molly.

"I brought an invitation home for all of us. It is a reception at the Waldorf for financiers, businessmen and their wives, daughters and sons. I do think you will enjoy yourselves. I could do without but I will take my girls to this, if they want," said Marcus.

Margaret spoke first, "Marcus, I think that would be wonderful. We seldom do anything together here in New York and I think you, too, would like to go, wouldn't you?"

"I would like to go," said Marcus.

"Well, I guess I would, too," said Madge.

When the Daly family entered the ballroom at the Waldorf they were immediately introduced into a group close by and offered champagne by a uniformed waiter. Madge and Molly had on pastel tea gowns made of silk and embroidered with flowers. Molly's was green and Madge's was yellow. They both were happy and at ease having their mother and father with them.

A young man came up to the group and said, "Mr. Daly, I don't know if you remember me, but I'm James Watson Gerard, serving in the New York legislature."

"Oh yes, Mr. Gerard, I do. I gave you a test case to solve that we had in Montana. I needed you in Montana, I said."

"That's right, Mr. Daly."

"Yes, Mr. Gerard, let me introduce my family. This is my wife, Mrs. Daly, and my daughters, Miss Madge Daly and Miss Molly Daly."

Mr. Gerard acknowledged the introductions and immediately started a conversation with Molly. "Miss Daly, what do you do with your time? I understand that you are not in school now," he said.

"I do some volunteer work with the Red Cross, and I play the piano and read a great deal," said Molly.

"Then you have definite ideas about world subjects, do you?" asked James Gerard.

"Well, I'm sure I'm not knowledgeable about everything in the world, but I do have opinions."

The conversation continued and Margaret noticed Molly getting agitated.

"What in the world do you think they are talking about? You had better find out, Marcus."

"All right. I'm sure it's fine, but I'll go over to them," said Marcus.

"Ah, there you are, Molly. Are you and Mr. Gerard enjoying yourselves?"

"Well, I don't know if enjoying is the word. We're arguing about the Spanish American War. I think we could have worked through diplomacy, and he says it was too late to even try diplomacy by the time we entered. And so he and I have differences. But it is stimulating, isn't it Mr. Gerard? I've never argued wars before," said Molly.

They all laughed at the idea of having such a conversation at a party, and Margaret was relieved to see them laugh.

Marcus, Madge and Margaret were escorted by one of the hostesses to another group, and Molly and Gerard moved into a different circle for conversation and introduction. Madge was with her mother and father when a young man was brought to them for an introduction.

"Mr. and Mrs. Daly, this is Henry Carroll Brown from Baltimore, Maryland. He is with his family's banking firm and is looking into an investment bank in New York. And this is Miss Madge Daly."

Henry Carroll Brown acknowledged the introductions and he and Marcus spoke about Anaconda and Montana while Madge and Margaret talked to each other.

Marcus suddenly realized that Carroll Brown kept watching Madge, and thought better of continuing the conversation, saying, "Mr. Brown, let's see if the ladies would like some refreshment."

Mr. Carroll Brown approached Madge and said, "Miss Daly, could I get you more punch or champagne?"

"Well yes, I do like the punch and it is warm tonight. I'll walk over to the punch bowl with you."

This was more than Carroll had hoped for. He had been watching this young woman all evening. He was so tired of all the "sweet young things" and saw that Madge was not only beautiful, but also serene and poised. She was so different. Carroll was intrigued.

Marcus said to Margaret, "Say, my girl, I think our two young ladies have found two young men that are very attentive."

"You noticed, too, did you Marcus? I'm pleased that those are the type of men attracted to our daughters. Let us hope our daughters are attracted, if not to them, to that type."

The evening carried on after dinner, and Mr. Brown and Mr. Gerard never left Madge's or Molly's side.

On the way home, Madge said, "Mama, Papa, I have never had a better evening in my life. Thank you for encouraging me to go with you."

"How about you, Molly?" said Marcus.

"Well, he certainly is a learned young man but very opinionated. He didn't let me win an argument, but then he never convinced me I was wrong. It was fun in a way. I guess I did enjoy myself," said Molly.

The year to follow found the two sisters dating and getting to know their new friends, James and Carroll. Marcus and Margaret realized that these relationships were getting serious. Harriot and Mark thought this courtship business was a bother. They couldn't be in the living room when the men came to call on Madge and Molly. They had to be quiet on the third floor and generally make themselves into perfect young adults or hear about it from their mother, Madge or Molly. Marcus refused to be involved, for he knew the dilemma of being in the way in his own home.

Another election for a senator from Montana was taking place and William Clark was at it again. He bought votes and won the election. Marcus, through his newspaper and detective work, found evidence to prove fraudulent election, so once again Clark's ambitions were stifled. It was a national scandal as well as the main topic of conversations for Butte residents and other Montanans.

On the fall return to New York, Marcus made a stop at Brooklyn. Sean Sullivan, the young son of the Sullivans that helped fifteen-year-old Marcus through customs, was a lawyer in Brooklyn. Marcus wanted Sean to consider going to work for Anaconda Copper Company. He knew Sean did some corporate work. He took a carriage to Sean's office in Brooklyn and entered unannounced into Sean's law firm. The male secretary was approached by Marcus.

"Wwhom do you wish to see, sir?" asked the secretary.

"I'm Marcus Daly, a friend of Sean Sullivan and his mother and father, and I would like to see him. I have not made an appointment, but am hopeful he can see me."

"Just a minute, Mr. Daly. I will find out if Mr. Sullivan is in."

The secretary returned and said that Mr. Sullivan would be pleased to see him.

"Sean, I haven't seen you in years, but I've kept up with you though your parents," said Marcus.

"Mr. Daly, this is an honor. I keep up on the latest of your successes," said Sean.

"Sean, I'll come right to the point. I need a lawyer in Anaconda that is familiar with corporate law. We have a lawsuit going that may be years in settling. Do you think you would like to go west? We pay top salary. There will be a house for you. I travel the distance between New York and Butte frequently, so you can come for visits here in your former home if you take this position in Butte. I'm going to be in New York for a while, but would like an answer as soon as possible — at least within two weeks," said Marcus.

"I'm sure I can know within this week. I need to speak to my wife and parents," said Sean.

"Of course, Sean, tell your mother and father 'hello' for me. They never would let me do anything for them and I'm always in their debt," said Marcus.

Sean accepted the position in Butte. He and his wife and young son settled in and Sean started delving into records, mine claims, court cases and the like, looking for a means to defeat Heinze. Sean was thirty-five years old, had black curly hair and blue eyes, and was tall and slim. His face was long, his nose was straight and narrow, and his white teeth had a gap between the front two teeth. He was a handsome mild-mannered man, but a tenacious fighter in court, as his previous cases attested.

The next year, 1897, found Marcus running between horse races on the coast, and business in Butte, Anaconda and Hamilton.

Marcus was glad he had Sean for he was finding him a great help in bringing information to Marcus. Clark had been defeated for his bid for a state Senator's seat, which made life worthwhile for Marcus.

Within a few days after the arrival in New York of the Dalys, Marcus kept an appointment with Standard Oil Company President, John McKay, and owner, John D. Rockefeller.

Marcus wasn't happy that Standard Oil Company was gaining control of his company, for there was no doubt that Anaconda was Marcus Daly's company. The partnership of Tevis, Haggin, Hearst and Daly had been successful. Marcus had complete power to operate the Anaconda as he saw fit. As an operator he had been a genius. Tevis and Hearst had already sold some shares to Standard Oil, and even Haggin was ready to release some of his. The company needed funds to operate, and going public was the quickest way to get funds unless they sold to Standard Oil Company. This meeting with Mr. Rockefeller and Mr. McKay would clarify what exactly Standard Oil had planned for the future of Anaconda Copper.

Mr. Rockefeller and Mr. McKay were seated and rose when Marcus entered the room. "Marcus, come take a seat, we will have a three-way conversation and will later iron out problems with our board and your partners," said Mr. McKay.

"Fine," said Marcus, "and I'll come right to the point. Do you want Standard Oil to own all of the stock? Do you want to buy my partner's stock and leave mine intact? I want to keep my stock."

"Now, now, Marcus, we want you to be head of Anaconda Copper Company, just as you are now, and you may also retain your stock, but we are giving two-for-one price on Anaconda stock. However, in your case and since you are Anaconda, we will have a legal document drawn up staing that your stock can be sold to Standard Oil Company at any time for the two-for-one price, the same as the rest of your partners -- or else sell at the current price of the day you decide to sell, whichever is larger."

"You see how very much we think of your expertise, Marcus. It will still be your company," said Mr. Rockefeller.

"It seems a fair arrangement," said Marcus, "and I will accept your offer. I know my partners think this is the right move; but, my friends, it breaks my heart to have Anaconda become a public company. For many years Anaconda has been mine and my people trust me and we have thrived together. The future could be difficult with the union bosses agitating for workers to join. Competition and law suits consume much time and money. I think you should delay public announcement of your intent until it is a fact."

Marcus wondered how much freedom he would have with Standard Oil as the majority stock holder. As he left the Standard Oil Company building he thought a walk would help clear his head and calm his feelings. As the meeting between the men had progressed, he had begun to realize that his company would be ruled by Standard Oil, even if he was Chief Executive Officer and President of Anaconda. This was not to Marcus's liking, but he had no choice if his partners sold their interest, which they intended to do.

The final proposition that John C. and Marcus finally agreed upon was that Marcus would retain his one-fourth interest in Anaconda Copper, and he would be the President and C.E.O. of the company.

Again he went over the meeting, and the one plus he could find in the whole matter was that he could relax more and the burden would be less on his shoulders. The decisions, the workers, the expansion, the town -- all had taken their toll on Marcus. He knew the chest pains were results of the stressful conditions of his position. Maybe, Marcus thought, this would be for the better.

"Well, Margaret, it is a fact. It has happened," said Marcus as he came through the hallway to her sitting room. He put the paper on her lap. The headlines read "Anaconda Copper Company bought out by Standard Oil." James Ryan would be Anaconda President, and Marcus Daly would be Chief Officer and President of the Board of Anaconda. The shares were going rapidly.

"Marcus, does it make you so unhappy? I should think you could be proud that the largest company in the U.S. wants to buy the company you developed."

"I guess it does bother me. I hate to lose the feeling that it's mine. Anaconda was always mine, even though I had partners. They even made me feel it was my own," said Marcus.

"Well my dear, maybe we can have more time together," said Margaret.

"I wish that were so, but I have a feeling that a few more years of putting out brush fires will be my lot in life. I've been all day at Standard Oil office. I've signed papers until my hand hurts."

"I wish I could comfort you more, Marcus, but you seem to be determined to mourn your loss," said Margaret.

"You're right, Margaret, I'll stop this nonsense. Actually, I'm twice as rich today as I was yesterday. I got $2.00 for $1.00 of my shares, and I still retained a one-fourth interest in Anaconda. I don't think they'll mess up my company," said Marcus.

Standard Oil bought the mine, but Heinze was still a thorn and would not settle without court cases, which he won. Each appeal delayed action longer.

Marcus spoke to Henry H. Rogers, chief executive of the Standard Oil Trust.

"Mr. Rogers, I have a young lawyer in Butte that has been researching the files and cases for information to discredit Heinze. His name is Sean Sullivan. I think you can use him to help in the battle," said Marcus.

"Well, Marcus, I'll have my lawyers contact him and see what he discovered and which direction to attack.

Chapter Eleven

The summer of 1899 at Riverside came and went. Molly and Madge were anxious to return to New York, for Mr. Brown and Mr. Gerard were there.

The horses were a source of pride for Marcus, and the trotters and racing thoroughbreds were winning and making history. In Hamilton, when the horses returned in their private railroad cars, the town would turn out to meet them, and carriages would take the trainers to their homes.

Christmas 1899 was a milestone. Mr. Brown and Mr. Gerard came Christmas Eve for dinner. After dinner the men went to the front living room for brandy and the usual business discussion. After the conversation ended about Standard Oil acquiring Anaconda, Carroll Brown spoke first.

"Mr. Daly, I have spoken to Madge and she gave me permission to ask you for her hand in marriage. She is a wonderful woman and I will do all I can to make her happy and you and Mrs. Daly proud."

"Does Mrs. Daly know?" said Marcus.

"Yes, Madge told her I was going to speak to you," said Carroll.

"Carroll, I would be honored to have you as a son-in-law and I know you will take care of Madge always."

Carroll took his handkerchief from his pocket and wiped his brow.

"Thank you, sir. I really have been worried about approaching you. I was afraid you would say 'no', and then what would I do?" said Carroll.

"Well, you needn't have worried. You know I think highly

of you. Let us drink a toast," said Marcus.

"Mr. Daly, sir, before you drink a toast I want to ask for Molly's hand in marriage," said Mr. Gerard.

Marcus was taken aback. "Men, two proposals in one night and to know that my 'yes' means my two daughters will be leaving is a lot for me to cope with," said Marcus.

"Well, Mr. Daly, Carroll and I thought we would both ask the same night. You know moral support means a lot. And it must be evident to you I think Molly the finest woman I've ever met. We argue, but it is really just discussions. We respect each other and I think we will have a fine life together," said James.

"I'm sure you will, James. If my Molly wants to marry you, I want her to have her wish. My blessings, boy. Now can we have a toast to our future together. Then let us go to see the women. They probably are waiting anxiously for us to come back to the parlor."

When they walked through the archway the women froze and looked at the men. Marcus went to Margaret, James to Molly and Carroll to Madge.

"Well, my darlin' girls, your young men have asked for your hands in marriage. I said 'yes' and your mother concurs. Right, Margaret?"

"Oh yes, Marcus. I'm so happy for you, Madge, and you, Molly, and James and Carroll. I'm proud to know you will be my sons-in-law."

Marcus called for Cutler to get a bottle of champagne and six glasses. "We are celebrating tonight," said Marcus.

The girls were in a daze, buying wedding gowns, preparing their trousseaus, and all the plans to be made. Even Harriot and

Mark were glad finally to be able to come and go as they pleased.

In January, Marcus had a cold and didn't feel well. He stayed in his hotel suite for a few days and didn't bother to notify the household. The bell hop and maids brought him whatever he needed. When Marcus lingered longer than a few days, the manager of the hotel called Mrs. Daly.

"Mrs. Daly, Mr. Daly has been in his room for five days and he doesn't seem to be improving. I thought you should know."

Thank you, Mr. Murphy. I thought Marcus had gone to the horse farm in Maryland. Please get a cab and have someone bring him home," said Margaret.

Margaret called the doctor and he arrived shortly thereafter. Marcus was telling everyone there was nothing the matter; he was fine. Margaret told the doctor what she had learned, and he went into Marcus' room to take care of him.

When the doctor came out he said, "Mrs. Daly, I think Mr. Daly is exhausted. Is there someplace he can go where he doesn't know anyone? Perhaps the spas in Carlsbad in Bohemia. That would be conducive to a peaceful life and Marcus could take the waters at the spring and heal. He seems to have an infection and I suspect it is his kidneys -- Brights Disease, perhaps. The Carlsbad springs are noted for their healing power."

"We will go immediately, doctor. I'll make reservations tomorrow. I'll take care of him. Thank you, doctor," said Margaret.

Margaret called the family and told them she was taking their father to Carlsbad. They would need to stay at the apartment and keep the household going. Harriot and Mark were instructed keep up with their school work.

Margaret called Standard Oil the next morning and told

them the plans for Marcus.

"Mrs. Daly," said Mr. Rodgers, "you must keep this out of the papers. It could start a sell-off of our stock. Marcus *is* Anaconda, you know."

"We're leaving today, so I don't think the newspapers will have a chance to know Marcus is ill," said Margaret.

"We are very anxious and we want you to know we care for Marcus. I'm sure he will get well as he is such a strong person," said Mr. Rodgers.

Margaret thought to herself. "Yes, Marcus is strong, but how much pressure can one person endure? Everyone depends on Marcus. He is like a horse that will die trying to please its master, not like a donkey who will lie down and not move if he's overloaded."

Margaret hurriedly packed and made arrangements to take the next boat to Europe, which left at 6:00 that night. Luckily there was a cancellation and she was able to book two staterooms with a connecting door. The ship doctor had been notified that Marcus was ill. Actually he had rallied during the night and his fever abated. Margaret was relieved.

Marcus was better, but he didn't have the strength to argue with Margaret. She was taking him to Carlsbad and they would stay there as long as needed.

The trip was uneventful across the Atlantic. Neither Margaret or Marcus felt a need to leave the staterooms except for short walks. Marcus was getting stronger, Margaret was happy. The circumstances weren't the best, but this was the first time they had been alone for years, maybe since Madge was born.

Marcus said, "Margaret, me darlin', where are you taking

me? Not that I mind. I'm feeling better all the time. Probably you are the best medicine for me. I've really enjoyed our being together on shipboard. I do feel relaxed."

"I'm so pleased that you're better, Marcus. I'm going to take care of you and protect you from yourself. It's time to slow down and enjoy life with me."

"Whatever you say. I'm still too weak to go against you, darlin'."

The ship docked at Le Harve, France, and arrangements were made to travel by train to Carlsbad.

Marcus and Margaret boarded the train. "Marcus, come sit here with me and look at the countryside. Isn't it beautiful?"

"Yes, Margaret, the beauty of mountains, valleys and lakes is always inspiring. It puts the human in perspective with nature. It is an awesome experience," said Marcus.

"My, you are philosophical, Marcus. I should have known that nature and vistas affected you, for you are a different person in the Bitterroot," said Margaret.

"What do you mean, 'different'? Could I love you more in the Bitterroot than on Fifth Avenue? I may get carried away with my projects and business, but it is you I love. Thank you, darlin', for rescuing me from myself and carrying me off on a second honeymoon. You did select a spa as far away as you could get. Probably it is for the best. I need to get completely away from business, or I'm back in the thick of it."

"I know. That's why we're going to Czechoslovakia. It's the most aristocratic spa in Europe. The accommodations are excellent and we'll be very comfortable. The waters are from 47 degrees to 165 degrees Fahrenheit, so you can bathe in the warm and drink from the cool. They are huge springs -- two million gallons a day is the flow of water."

"Good heavens, darlin', I'll be completely saturated in and out. Never heard of so much water bubbling from the earth."

Carlsbad was a lovely as any place Margaret and Marcus had been. They were very pleased. Even though Marcus was recuperating, they were able to enjoy the village and excellent food. It had been years since they had walked on paths, and they did do so instead of riding horses or taking carriages. The walking strengthened Marcus, and Margaret thought her figure was slimmer.

"Margaret, do you think we've been here long enough? I feel very fit now and I promise not to let myself overdo in work or play. I'm turning over a new leaf. We've been here two months and it's time to go to Riverside. You know I relax more there. What do you think, Margaret?"

"If you are ready and feel well again, we could leave. I'll wire the household to ready themselves for our arrival in New York within three weeks, and to prepare for the journey to the Bitterroot."

"Wonderful, my girl, I've been idle long enough. Don't look at me like that. I told you I would take care."

When Margaret and Marcus arrived at 725 Fifth Avenue. Molly, Madge and Harriot rushed to the arms of their mother and father.

"Oh, Papa, we are so glad you are well," said Molly. "We've been worried. You're never sick."

"Mama, we missed you so. That is the first time we've been alone. Harriot and Mark were great help and we did have some good times," said Madge.

"Mr. Gerard and Mr. Brown have made daily calls and we're ready to set our wedding dates. We don't want a double wedding," said Molly.

"Whoa, me darlin's, let me make my way through the hallway before we solve all the problems of our lives," Marcus said as he laughed at the four who were all talking at once. Harriot took her Father's hand and walked with him into the living room. Mark walked on the other side of Marcus.

"Papa, will you be able to ride with me this summer?" asked Harriot.

"Of course, girl. You think your Papa is an invalid? I feel fine, and better every day. Your mother was as good as medicine for me, along with the famous waters of Carlsbad."

"Mark, how is school? Did you get your application into Yale?"

"School has been fine this year and I'm sure to be accepted at Yale. I'm at the top of my class and already I have a roommate, Cornelius Vanderbilt. We've become good friends, and we're going to try out for the rowing team at Yale," said Mark.

"Harriot, in two years, I'll be crossing the ballroom with you for your debut. My girl will be the most beautiful and talented debutante of the year 1902!" said Marcus. Tonight I would like for you to play the piano. I've missed your playing for me when we were in Bohemian country."

"Of course, Papa. I've learned some beautiful new classics that I think you will enjoy," said Harriot.

Soon after the arrival of Marcus and Margaret, the family was ready to leave for Riverside.

Marcus made a brief trip to Standard Oil to see Mr. Rodgers.

"Well, Marcus, you look fit again. I'm so pleased. But do take it easy and let us handle that problem with Heinze. We can spare you that ridiculous fiasco," said Mr. Rodgers. "I guess you would recommend the waters at Carlsbad, eh, Marcus?"

"Yes, I would and I'm feeling fine now. I even lost a little weight, which can't be bad for me. I may need a tailor to take in an inch or two on my suits. I'll be stopping in Butte for a day or two and Anaconda for a day or two, but I'll take it slow. We have good men in charge."

"That reminds me, John Ryan reports that your young man, Sean Sullivan, is a great help to him. Thought you would like to know that," said Mr. Rodgers.

"'Tis good to hear a fellow Irishman is making his way through the maze of the business world. Thank you for telling me. I'll go now, as Margaret frets if I don't return at the designated time."

The men shook hands and bid each other goodbye. Marcus promised to be back in the fall with a full report and recommendations, if needed, for Anaconda Copper Company and the Anaconda Smelter.

"Good Luck, Marcus, I'll see you in September," said Mr. Rodgers.

This was the thirteenth year that the Daly family had summered in the Bitterroot Valley at the Riverside estate. The routine was the same: the summer help came year after year and the stand-bys, Mr. and Mrs. Cutler, Stena and Eddie, were in charge. The children had grown into adults and young adults. Their interests changed, but some interests remained constant. The summer at Riverside was a blessed part of each of their lives.

Marcus went to the stables as soon as possible and sought out Sam.

"Where are you, Sam?"

"Here, Marcus, in the tack room going through the files," said Sam. "How are you, Marcus? You look fine. We were all concerned to hear you'd been struck down. Looks as if you beat the ailment."

"I did, Sam, and I'm fine. What's new here? Where is Tammany? I see Montana is in his stall. Are the rest, including Tammany, in the pasture across the way?"

"Yes, they are. Montana is kicking and biting, so we're going to take him farther down the lane to another pasture."

"Speaking of pastures, the Henry Cross paintings of the horses in the pasture are very good. We could identify each horse. Come in the tack room. I've got those big ones of Cross' in there. It was some job to get them hung!" said Sam.

Marcus had not seen the paintings hung in the tack room, and he approved.

"Mr. Daly, I believe Hamburg may be our greatest stallion. He has performed well on the track and at stud. You should have winning horses for many years. The foals from him are doing well and will be ready to race in 1902."

"I'm happy Sam. You've done it all with no supervision. It's wonderful to have a man that can carry out my plans without giving me any need to worry. Come, let's go to the pasture."

As they approached the pasture, Marcus called to Tammany. He came trotting over, head high with the sun glistening on his light chestnut coat.

"You old man, Tammany. Come see me and tell me all about it."

Tammany whinnied, shook his head and responded to Marcus rubbing his mane and stroking his back.

"We're old friends, aren't we, Tammany? You are my greatest horse. The others may win more and sire more colts, but you are the one who made the Daly stable recognized worldwide."

Tammany responded as if he understood and moved closer to Marcus.

"Well, Tammany, I must go, but I'll be back to see you," said Marcus.

"Say, Sam, how about having my three top stallions do a little walk around the track during the final race at the fair? I think the audience would like to see $100,000 worth of horseflesh. Tammany, Prodigal and Hamburg are the foundation of the Daly racing stable."

"Great idea, Marcus. The crowd will love it. Are you going to turn into a showman as well?"

Both Sam and Marcus laughed, for who knew what Marcus might do next?

Marcus wanted to check the orchards, but he would do that another day. He had had enough outdoor activity for the day. Besides, it was time for family.

The valley was shipping apples around the state and some out of state. It had become quite an industry in the valley, and the Riverside vegetable and fruit farm had prize apples to sell. That was proven at the Ravalli Fair. The Riverside McIntosh apples were judged as the finest. The pears were doing well, but the weather made it more precarious for pears and peaches to mature.

People were moving to the Bitterroot, buying a few acres and getting into the orchard business.

Marcus inspected the orchard and the sawmill, and met with Ravalli County Bank president, Mr. McCrackin, and with others in Hamilton. If any business, school or church needed money, they knew Marcus would help them. Hamilton and its programs were a part of his overall plan for excellence, and the population were the beneficiaries.

The fair was busy, with 10,000 people attending.

The last day of the fair and the last race, Marcus was in his box with the family. "I've got a little surprise for you all," said Marcus. "Watch the track before this last race coming up."

Announcements from the track came across the loud speaker. "Ladies and gentlemen, we've had a grand time at the fair and the races have been spectacular. We thought it only fitting that you see our own Bitterroot products of Marcus Daly's stable. I present to you the greatest stallions who will influence the outcome of racing for generations. $100,000 is the value placed on these three -- Tammany, Prodigal and Hamburg!"

The horses had garlands of flowers around their necks, and the young black jockeys riding the horses were wearing the green and gold silk uniforms of the Daly stables. The crowd rose to their feet and gave a standing ovation for these famous horses. The announcer interrupted the applause saying, "And here in the stands is Marcus Daly and his family. Mr. Daly has brought fame to the Bitterroot in many ways, but today we honor you, Mr. Daly, for the enjoyment and pleasure and fame that your thoroughbred stable has brought to us. Thank you."

The crowd again broke into applause and stood to show their respect to Marcus Daly and his family. Marcus, Margaret and the children were urged to stand and again the applause was thunderous.

"My gad, I didn't expect that," said Marcus. "I just thought the people would enjoy seeing the three famous stallions."

"Well, we all survived the attention and it was not unpleasant, just unusual," said Margaret.

The last race was anticlimactic after the ringside show. The horses ran well and the winners collected their money.

When the family and guests had finished dinner that evening, after a full day at the races and fair, Marcus excused himself from the group and went into his office. There was a cylinder addressed to Mr. and Mrs. Marcus Daly from Mr. Gibbons, the architect in Missoula. Marcus opened the cylinder and drew out the blue prints for redoing the Queen Anne Victorian mansion.

Marcus couldn't believe it was the same house, but it was. The turrets were removed, and it had a new exterior of buff-colored brick and an edifice of classical columns and pediments. It reminded Marcus of London, as well it should. The latest architecture being used for homes, mansion and public buildings was influenced by the Georgian Revival period.

Just then Margaret knocked at the door. "Marcus, aren't you exhausted? What are you doing? May I come in?"

"Of course, darlin'. I'm looking at the blue prints for the remodeling of this 'church'."

"Now, Marcus, it does not look like a church."

"I'm just foolin' you, Margaret, but I do like this much

better. I liked Regents street in London and of course it's achitecture is from the original Georgian Period. Am I right, darlin'? See, old Marcus does observe and learn!" said Marcus.

"Marcus, this is grand. The inside rooms have been enlarged and we would have three big rooms and a large front foyer. That would be much better, since we're always separated after dinner because we are too many for any one room. But, Marcus, it is such an extravagance, for it has only been a short time since we did the Queen Anne."

"Don't worry, Margaret, we can do it. Besides, since we spend more time here together, this is our real home!" said Marcus. "We'll do it next spring and have the mansion ready for us. I'll write Mr. Gibson and tell him we accept his concept."

Back in New York with the young people in school, Molly and Madge being courted by Mr. Gerard and Mr. Brown, Marcus and Margaret were at peace with the world and each other.

"Marcus, don't you think you want to move back from the Nederland Hotel and be with your family?" asked Margaret.

"Yes darlin', I do but let us wait until the girls are married. It will be a busy time for you, and I have some business and loose ends to clear. As soon as the girls are married, I'll be here. It will be our time at last. I'll give up all this traveling except for going to Riverside or other needs of the family. I've missed a lot of the family traditions, but I don't intend to miss anymore. Mark starts at Yale and I want to take part in the activities for parents."

So it was decided that after the weddings Marcus would be at home at 725 Fifth Avenue.

Margaret was busy with Molly and Madge. They might as well have had a double wedding — it was double trousseau and double expense, Margaret thought. But they each wanted their day and she could understand that.

Margaret called to Molly and Madge. "Your father is coming for dinner tonight to be with your fiancés, and Mark and Harriot will join us. It will be a family party."

"Wonderful, Mama. Carroll and James will be pleased. They enjoy Papa and his stories. Papa's life is so different from any other life they've known," said Madge. Molly echoed her pleasure at the surprise for the evening. Molly loved a surprise always.

Margaret heard the doorbell. Cutler came into the room.

"Yes, Cutler, what is it?" asked Margaret.

"It is the assistance manager at the Nederland. Mr. Daly fell on a grate outside the hotel. He was going to walk over here. Mr. Daly seems to be in pain and they have him in his suite at the Nederland."

Margaret's heart stopped. "Cutler, get the carriage, we must go and bring Marcus home. Tell Mrs. Cutler to prepare Mr. Daly's room. Hurry, Cutler."

Margaret and Cutler arrived at the Nederland. The hotel employees were wringing their hands, wondering what they could do. Margaret entered Marcus' suite.

"Marcus, my dear, we're here to take you home. Mrs. Cutler has your room ready. Oh dear me, are you able to be moved?" asked Margaret.

"Yes, darlin', I want to come home. Some of these big young men will carry me to the carriage and come along and help me into my room. Won't you, boys?" asked Marcus.

"Oh yes, Mr. Daly, let us help you."

"I'll be all right, lads, don't fret so. You look positively depressing.

I'm fine -- just a bad fall and my old legs are wobbly," said Marcus.

Finally Marcus was settled in his room. Mr. Brown and Mr. Gerard had arrived, for there had been no time to notify them of the catastrophe.

Marcus wanted only a little soup and some soda crackers. That worried Margaret, for Marcus was always ready for his dinner.

The doctor came and examined Marcus. "Mrs. Daly," said the doctor, "I think Marcus had a fall that injured him internally or else it bruised his kidney, for I see the same signs as he had before he went to Carlsbad for the water cure."

"Oh, doctor, what can we do for him?" asked Margaret.

"There isn't much, but do keep liquids down him. Maybe we can flush out the poison," said the doctor. "He does need rest and quiet. See that he sits up each day. I will be here every day to watch his progress."

Margaret went to the family in the living room and told them what the doctor had said.

Margaret said, "We'll know in a few days if he's healing. In the meantime we will keep him company and make him as comfortable as possible."

"Poor Papa," said Harriot. "He gives all, asks for nothing and then has to endure this pain. It isn't fair." Margaret was startled by her analysis. It was true, but she didn't think Harriot would be that observant.

The next week was a trial for all. Marcus rallied, then he would sink into weakness again. Marcus called to Margaret, "Come sit by me, darlin'. I'm afraid we may not have our time together as we hoped and planned. The pain is bad and I try very hard to ignore it, but there is something wrong. My one thought is for you and the family. Margaret, I'm grateful that if one of us must go first, that it is me. I know you can go on with life, Margaret, but if you were to go, I couldn't go on."

"Now, Marcus, don't talk that way. We're going to get you well." The tears brimmed over and fell down Margaret's cheeks in a rush.

Margaret told the children that their father was failing and was giving up hope. Margaret asked each of them to talk to him for short periods and try to make him happy. It was a hard task for each of them.

Madge and Molly chose to visit their father together. "Hello, Papa. We've brought you some fresh orange juice and some sugar cookies. Would you like one of us to read to you?"

"No darlin's, but I would like to see your young men. I know you had to postpone your wedding, Madge, but I will make it up to you if I get out of this bed," said Marcus.

"Molly, I should be well enough for your wedding, darlin'."

The young men came to see Marcus the next day.

They were startled by Marcus' color, for he was jaundiced and looked as if he were in pain.

Marcus brightened when he saw Carroll and James. "Good of you to come. I want to tell you that you men will have to help my family. I don't think I'm going to be able to beat this dastardly disease. I want you to know that I am very happy to have you as my sons-in-law and am only sorry I probably won't be able to enjoy life with you in my family."

James said, "Mr. Daly, you've just had a setback — we'll be out at the races next spring."

"That's right, Mr. Daly," said, Carroll. "Madge and I will meet you at your horse farm when the horses come in from the Bitterroot for the racing season."

Marcus got a sparkle in his eye when he heard spring and horse racing, but it soon faded as another knife-like pain pierced the steady pain of the failing kidney.

Harriot took her turn with her father. Marcus said, "Harriot, I can hear the piano from here. Would you go and play for me?"

"Certainly, Papa, I would love to. I'll always play for you. I can't wait for you to get well. I'm going to give a concert during the Christmas holidays and I want you to be there for me."

"I will darlin', and we'll ride in the Bitterroot next summer. Your Shamrock is still a fine-looking horse, and you never looked more beautiful than when you and Shamrock are crossing the countryside at a canter."

Harriot played three pieces and went back to see her Papa, but he had fallen asleep.

Mark brought his father the morning paper the next day. Marcus had had a decent night and looked more rested.

"Papa, you look better today. Maybe the crisis has passed. I brought you the paper. Standard Oil and Anaconda are in a write-up. Just says that the stock has risen since the purchase of Anaconda."

"And well it would," said Marcus. "There are millions of dollars still underground in that copper ore!

"Mark, if I don't pull through this illness, you must help

your mother. I'm glad you are majoring in economics, and I know you will apply your good common sense to help preserve what fortune I've accumulated. The will has been drawn up and is in the bank box."

"Papa, please don't say you aren't getting well. You look much better."

"Gad, boy, you know I want to get well, but if I don't I want you to be aware of your responsibility. James and Carroll will help you, too, but you are blood relation and should have the responsibility for final decisions."

This talk with Mark had stirred Marcus, and by the time Margaret came in, Marcus looked ready to get up and walk out.

"Marcus, you look so much better." said Margaret.

"I feel better too, but I have something that has been bothering me. I haven't been a good Catholic as far as going to Mass, but I do believe in God, and I want a Catholic burial. Please call a priest. If I confess my sins now I will be ready to go, or I will revive and live my life with you and the family."

Margaret called for a priest for St. Patrick's Cathedral. Father Tom came immediately and went into Marcus with the communion wafer, intending to give him last rites after Marcus' confession. Marcus had lost his vigor and was again lying quiet with his eyes closed. He opened them to see the priest.

"Well, Father, I guess you have time to hear my sins. It's been about thirty-five years since I've been to confession." After Marcus made his confession and received last rites, Margaret came to be with him in his last hours.

Margaret didn't realize that Marcus was leaving her. As she held his hand and talked, she felt his hand jerk and looked at

Marcus. He was dead.

Margaret was stunned. How could this be? she thought. Marcus the vigorous one.

"Marcus," Margaret called, but he was gone forever. Margaret thought, no more sweet kisses. I can't bear it!

Margaret lay across Marcus' chest and stroked his face. My heart has been stabbed, she thought. How can I tell the children?

In her sorrow, Margaret, heard the words Marcus had said, "You can go on with life, Margaret. If you were to go, I couldn't go on."

Margaret opened the bedroom door into her room and rang for Cutler.

Cutler answered the call and entered Mrs. Daly's bedroom.

"Cutler, please call the family to the living room. Mr. Daly has died."

"Oh, madam, what a terrible loss. We all loved Mr. Daly. I will go and tell the children to come to the living room."

Margaret was dry-eyed and wondered what was carrying her down the hallway to the living room. There with all the children looking at her, she said, "Your good father, my beloved husband, is dead. He died a few minutes ago. He just fell asleep. We will have a Catholic funeral. Those were his wishes."

The news spread rapidly. All the Catholic churches in Montana synchronized their Masses with the time of Marcus Daly's funeral Mass at St. Patrick's Cathedral, and the bells tolled throughout Montana's towns on the appointed hour.

The family sat in an alcove, shielding their grief from the rest who had came to pay their respects to Marcus Daly. The mourners were from all walks of life, tycoons, gamblers, servants, waiters, jockeys.

When the Mass was over, Margaret, Madge, Carroll, Molly, James, Harriot and Mark left the church and proceeded to the family mausoleum where Marcus would be the first of the family to occupy the crypt.

After the funeral, after the children had retired, after the lights had gone out, Margaret lay in her bed.

"Will I ever be able to sleep again? Will I ever be able to eat again? Will I ever be able to comfort my children when I can find no comfort? I can't imagine life without you, Marcus." Just then a tingling on her lips and a pressure touched them. Am I losing my mind? Margaret thought. Maybe it was a last kiss from Marcus. It was a comfort and she cried herself to sleep, awoke, and then slept again.

When Margaret was aware that morning had come ,she knew that Marcus was with her, inside her forever.

She knew then that she could face the sorrow, make the necessary decisions and care for her children as always. She touched her lips and said to herself, I know you are here, Marcus, here with me!

Epilogue

There was complete silence as the tour group looked at the portrait of Marcus Daly, and knowing that Marcus had died at the age of fifty-eight in November 1900 in New York City.

The guide broke the silence, saying, "Isn't it amazing what Marcus Daly accomplished in the short number years he lived? Especially the last 15 years of his life when the great wealth from the copper mines and smelters made him one of the wealthiest men in the United States and the world.

"Now of course his widow would be a woman of means, and a lavish existence for Margaret Daly and her four children was guaranteed.

"Do come with me to the sliding doors and look at the most important room in the house. The children all played the piano, many nights of music were enjoyed during the summer months in this music room with its Steinway grand piano. The gold damask wall glowed in the light of the tall lamps with the beaded shades, and the sconces with lights on the walls.

"Here is a portrait of Harriot, who married Count Sigray from Hungary. She met him at the wedding of her friend the Vanderbilt daughter, who became Countess Szechenyi. Count Sigray was best man for Count Szechenyi, and Harriot caught his eye. They married and went to live in Hungary and had one child, Margit (Countess Bessenyey).

"Through this sliding door is the living room, done in greens and golds with comfortable green velvet couches, large armchairs and a fireplace with a conversation grouping of two love seats. This was where the adults, who were dressed formally, gathered before dinner to have a cocktail, fruit punch or tea before

7 p.m., at which time the group made their way to the dining room for a lavish feast. The children went to their own dining room at 6:00 with their nurse and nannies, until they were eighteen years old. After turning eighteen, the young joined their elders for dinner in the formal dining room.

"After Mr. Daly's death, Margaret wondered if Riverside would ever be the same. The summer after his death she longed for Riverside and invited all of her friends and relatives to come for the summer, and they did.

"Arriving the first summer after Marcus Daly's death was the oldest daughter, Madge, with her new husband, Henry Carroll Brown, a Baltimore banker. Molly, the second daughter, arrived with her new husband, James Watson Gerard, who was a member of the New York State Supreme Court, later to be the U.S. Ambassador to Germany during World War I. Mr. and Mrs. Lewis from California, Mr. and Mrs. Evans, sister Mattie and her nursing companion, and Dr. Murray and other family and friends came filling the house. Harriot and Mark helped their mother with the entertainment of guests and family. Margaret Daly found that there was a life at Riverside even after her beloved Marcus, the founder of Riverside, was gone.

"And so Margaret was hostess at Riverside from 1887 until her death in 1941. She died quietly in her sleep in her bedroom in the Daly Mansion, attended by her daughter Molly, her nurse, Virginia Rosco, and her grandson, Marcus III (Mark's son by his marriage to Ellie, a woman from Vienna whom he later divorced. In turn, Mark was granted custody of Marcus III).

Marcus II preceded his Mother. He was a great sportsman and hunter and added the trophy room to the Mansion in 1912. On an annual bird hunting trip in Virginia with his friends, he fell behind the group; when they sought him out, they found him dead of a massive heart attack. He was 47 years old.

"Margaret Daly had known great sorrow when her oldest daughter, Madge, died at age 36 from a heart attack, leaving two little girls, three and five years old, to be reared by their father. Their inheritance was guarded by their grandmother Daly. Since Margaret's only son, Mark (Marcus), had been taken from her, she also became the guardian of Marcus III. He was ten years old and she was seventy when he became her charge.

"Riverside was the place for renewal of friendship, love, and bonding between relatives and friends for fifty-four years. Mrs. Margaret Daly died in July of 1941 at Riverside in Hamilton, Montana. World War II was declared on December 7, 1941. After Margaret Daly's death, the house was cared for by Eddie. The ranching continued; the war took its toll. After two years, Eddie decided no one was coming to Riverside. He boarded the mansion windows and left without anyone knowing why or where.

"Harriot's beloved Count Sigray was cast into a concentration camp by the Nazis. Harriot and her daughter Margit were held under house arrest in Hungary on the Sigray Estate until the end of World War II, 1945. Count Sigray died one year after his release from the concentration camp, and Harriot died one year later. Both deaths were the result of harsh treatment and malnutrition and illness from World War II.

"Margit, the daughter, had returned to the United States after World War II with her mother and father. After their deaths she used her wealth from her inheritance to buy out the interests in the Daly Estates from her two cousins in Baltimore (Madge's daughters) and her cousin Marcus III. Margit then became owner of five estates: one in California, another in Baltimore, a Maryland farm, the New York Fifth Avenue Apartment, Riverside and the Daly Mansion. She then married Baron Bessenyey, who had three children, and who died one year after their marriage.

Margit never opened the mansion, but she did turn Riverside into a cattle ranch and a Hungarian horse ranch (Hungarian and Arabian horse cross-bred). She was a famous horsewoman and lover of animals.

At Margit's death in 1985, her stepchildren inherited her estate. Unfortunately for the heirs, Montana law requires non-blood relatives to pay a large inheritance tax.

To relieve this financial burden, Francis Bessenyey, the executor of the estate, offered the State of Montana the Daly mansion and fifty surrounding acres in exchange for $400,000 of tax relief. As a result, the Daly mansion was finally opened to the public after 41 years of darkness and quiet."

The guide continued her narrative, "Oh, here we are in the children's dining room. Isn't this lovely wallpaper? The children loved the birds with bright plumage. Notice how lovely the wallpaper is throughout the house. It was all imported from England for the 1910 Georgian Revival renovation that Mr. Gibson had drawn up for Mr. and Mrs. Daly. Since Marcus had preferred the Georgian Revival to the Victorian, it became a tribute to Marcus from Margaret, with love."

The guide turned to her group and said, "This is the end of the tour. Through this door into the hall you will exit at the North Portico. This was the family's entrance into the house. Only guests came through the front door. Thank you for coming. Please come back to the Daly Mansion again and again."

The tour guests went on their way after thanking the tour guide. Each guest also stopped at the gate to tell the gatekeeper that she was right -- it was a magnificent house and a marvelous story of the Daly family and of the man Marcus Daly.

BIBLIOGRAPHY:

I. Books:

Montana - An Uncommon Land	K. Ross Toole
The War of the Copper Kings	C.B. Glasscock
Queen of the Golden Age	Cornelius Vanderbilt, Jr.
Encyclopedia America	1953 Edition

II. Magazines:

Town & Country - "Forever Fifth" Lindsay Van Gelder
Town & Country - "The Horse of the Nineties" Wendy Insinger

III. Newspaper articles from Georgiann Dayton Collection:

New York Times....................................New York City, NY
Anaconda Standard................................Anaconda, Montana
Ravalli Republic.....................................Hamilton, Montana
Missoulian...Missoula, Montana
Montana Standard.................................Butte, Montana